Integrated Science

Volume 18

The **Integrated Science** Series aims to publish the most relevant and novel research in all areas of Formal Sciences, Physical and Chemical Sciences, Biological Sciences, Medical Sciences, and Social Sciences. We are especially focused on the research involving the integration of two of more academic fields offering an innovative view, which is one of the main focuses of Universal Scientific Education and Research Network (USERN), science without borders.

Integrated Science is committed to upholding the integrity of the scientific record and will follow the Committee on Publication Ethics (COPE) guidelines on how to deal with potential acts of misconduct and correcting the literature.

Stefania Achella · Chantal Marazia
Editors

Vulnerabilities

Rethinking Medicine Rights and Humanities in Post-pandemic

Springer

Editors
Stefania Achella
University of Chieti-Pescara
Chieti, Italy

Chantal Marazia
Heinrich-Heine-University Düsseldorf
Düsseldorf, Germany

ISSN 2662-9461 ISSN 2662-947X (electronic)
Integrated Science
ISBN 978-3-031-39377-8 ISBN 978-3-031-39378-5 (eBook)
https://doi.org/10.1007/978-3-031-39378-5

This Springer imprint is published by the registered company Springer Nature Switzerland AG
The registered company address is: Gewerbestrasse 11, 6330 Cham, Switzerland

Introduction

This initial fragment of the millennium has confronted us with the conjuncture of portentous calamities, natural and manmade. The conjunction of the 2008–2009 financial crisis and its persisting effects, with the rapid succession of pandemic events culminating in the COVID-19 global emergency [1], has impacted on a growing sensitivity towards momentous challenges to the human species as a whole, such as climate change and migratory crises, conjuring the perception of an inexorable loss of control of the human being over self and the world. The Western Promethean idea of the self-made man has been undermined by the growing awareness of the limits of human beings and of the risks to which not only the most fragile, but all living things—human and otherwise—are exposed.

Kate Brown's diagnosis of a "vulnerability Zeitgeist" [2][1] of British social policy can be profitably extended to cover the acknowledgement of such precarious condition, inviting reflection on its manifold facets and stimulating a shift in the perception of the "modern Man", from an autonomous and responsible—male, white, Western—agent to a "vulnerable victim who must be protected and enabled" [3, p. 360]. Since the 1980s, the fragile and vulnerable nature of human and non-human living beings has grown into a useful interpretative tool: from the capabilities approach to the concept of *care*, from recognition theory to post-structuralist currents, vulnerability has gained increasing ground in philosophy. The concept has also found application in sociology (think of the studies of Marie Garrau and Robert Castel) and in the field of rights (Martha Fineman's "vulnerable subject" in the legal field [4]). In 2014, Laurence Burgorgue-Larsen waved these different interdisciplinary threads in the term "vulnerability turn" [5, p. 240].

So intended, the concept of vulnerability entailed reconsideration of obligations at the inter-subjective, as well as at the state-level. Feminist thought has contributed

[1] Brown speaks about "vulnerability *Zeitgeist*", explaining that despite different opinions, vulnerability has become a popular conceptual mechanism.

to making vulnerability a central descriptive, but also prescriptive category.[2] Feminist movements[3] were the first to articulate that vulnerability does not only refer to a class of subjects placed under tutelage.[4] On the one hand, they have stressed its universality; on the other hand, they have shown how vulnerability is not independent of external conditions, social, political and cultural. Hence, the shift from the descriptive to the prescriptive, normative dimension: to the plainly descriptive dimension (who is vulnerable), a political layer has been added, forcing us to interrogate the why of vulnerability. This discourse, extending beyond gender, has marked the field of theory, epistemology and practice over the last 30 years, opposing the hegemonic idealisation of the unattainable invulnerability of the individualised modern liberal subject, as well as the dichotomous distinction between agency and vulnerability. This "theor-ethical stance" is not devoid of limits: from the risk of a compassionate approach towards the most vulnerable, to an ontological reductionism possibly resulting in essentialism, and stigmatisation of differences, and restoration of paternalistic practices.[5]

The recent pandemic has even more vividly revealed the pitfalls inherent in the concept of vulnerability, be it understood as a characteristic of all living things, or as an attribute of categories of people, classes or individuals. In the midst of these difficulties, the need for a radical change of perspective, in order to face such extreme (and increasingly frequent) unbalances, has emerged with unprecedented clarity.[6] Institutions and organisations have striven to imagine "a future world that is more 'equitable', 'sustainable', 'resilient', and one that pays more attention to social

[2] Recent years have witnessed numerous declinations of the concept: "social vulnerability", "ontological vulnerability", "intrinsic vulnerability", "situational" and "pathogenic vulnerability", "racial vulnerability", "economic vulnerability", "epistemic vulnerability", "teleological vulnerability", "embodied vulnerability" and "moral vulnerability". Vulnerability can also serve to qualify and refine central concepts, from judgement (e.g. Kimberly Hutchings' "vulnerable judgement") to subjectivity (e.g. Fineman's "vulnerable subject" instead of "liberal subject") [see 6]. Each of these declinations carries a descriptive value of the type of vulnerability characterising the human, as well as an expression of how it should be understood and how to mitigate the risks of this essential characteristic.

[3] As Polychroniou notes, pointing out the prominent role of feminism in the development of a theory of vulnerability, "the recognition of our constitutive vulnerability inaugurates an alternative philosophical paradigm beyond social contractualism by critically problematizing the archetypical emergence of an apriorically male, adult, independent, rational, self-interested and self-sustained subject, which equally, freely and consensually forms mutual contractual bonds and social relationships with others" [7, p. 115].

[4] Consider in particular the application of this concept in the fields of care ethics, feminist phenomenology and jurisprudence, gender violence studies, eco-feminism, as well as in feminist epistemology and post-humanism [e.g. 8, 9, 10].

[5] Some critics have even suggested that this term be dismissed due to its vagueness. According to Wrigley, for instance, the term would, if anything, act as a linguistic device to draw attention to certain problems, but less effectively than other clearer and more precise concepts [11].

[6] "The solution to the health, economic, and social crises COVID-19 has precipitated", reads a volume written during the pandemic, "is not to become more insular. Rather, it is to recognize that our future is co-determined with others, and to seize the benefits of our interconnectedness within and across countries" [12, p. 27].

'well-being and inclusiveness'" [13, p. 2; see 14]. However, "how these goals can be achieved, if they are possible at all, [remains] subject to debate, controversy, and conspiracy" [15, p. 4].

In the present volume, we have attempted to put the "vulnerability" paradigm to the test in its manifold declinations—hermeneutical, theoretical, critical and operational—trying to understand how it can branch in the sphere of rights, justice, equity and with what effects. The chapters bring together different perspectives, disciplinarily and methodologically. The interpretations of the pandemic event with regard to possible future implications provide here both a hermeneutic barycentre and an access point to wider horizons. Some authors identify this global emergency as an opportunity to overcome distortions of a social, political and anthropological vision built on the marginalisation of individual or collective fragility; others, instead, have underlined how the pandemic has displayed practices that, rather than addressing fragility, have accentuated differences and discrimination. Believing that the complexity of the issue requires a dialogue among divergent opinions and perspectives, we have articulated the problem of vulnerability in three different levels, mirrored by the three parts of the volume.

Part I, *What Is Vulnerability?*, collects reflections on the meaning of vulnerability in its historical (Achella), biopolitical (Fulco), ethical (Donatelli), psychological (Donise) and biosocial (Dragano) dimensions. The first essay, "Vulnerability is Said in Many Ways" by Stefania Achella, analyses the main lines of the debate around the concept, highlighting the aporias stemming from its use. Achella tackles the question of how "vulnerability" can provide an ethical and political perspective for the future. Rita Fulco's contribution, "Humanity of the Human and the Politics of Vulnerability", focuses on the relationship between the horizon of politics, institutions and practices and the more ontological one of the definition of the "humanity of the human". Before analysing the political discourse on vulnerability, Fulco engages with two of the main hermeneutic models of recent modernity, those of Judith Butler and Simone Weil. She thus situates vulnerability at the crossroads between the constitutive dimension of the human and its socio-political nature, connected to discourses of power. Piergiorgio Donatelli's "Ethics in Scenes of Disaster" points at the importance of shifting attention to interdependencies, drawing from anthropologist Veena Das' analyses of disasters and on Ludwig Wittgenstein's and Stanley Cavell's reflections on the ordinary. Taking the move from the concept of vulnerability, Donatelli shows how the COVID-19 pandemic—like other major disasters we have witnessed more and more closely in recent years—has proven the importance of care work, carried out by many people who are usually invisible, but crucial in times of crisis. In her essay, "Vulnerability and the End of the World. Trying to Read the Post-pandemic Age (with Karl Jaspers and Ernesto De Martino)", Anna Donise explores the idea that preventive measures and restrictions imposed during the pandemic brought to light the existential and psychological discomforts associated with coping with vulnerability, leading to a kind of collective delirium. By placing the thoughts of psychopathologist Karl Jaspers and ethno-anthropologist Ernesto De Martino at the centre of her analysis, Donise uses the category of the "end-of-the-world" to describe distressing experiences and to understand how they can be incorporated into a comprehensive concept of community through cultural understanding. Nico Dragano's chapter, "A Biosocial Perspective on (COVID-19) Pandemic Outbreaks: Interfaces of Biology and Social Determinants", closes this first part. Dragano shows how vulnerabilities, in cases such as pandemic spreads, are at once biological—the increased biological susceptibility of a person to becoming infected with a specific pathogen—and social—the cases of socially disadvantaged groups. Dragano points to the strong interaction between these two levels and, consequently, to the need to develop a multidimensional conception of vulnerability, which he accomplishes by linking the classical model of infectious disease epidemiology with ecosocial theory.

Part II of the book, *Who Is Vulnerable?*, provides a closer focus on subjects and classes considered vulnerable: women, people with mental illness and/or disabilities, children and people confined in institutions such as prisons or long-term facilities. In her essay "Vulnerability and Gender After COVID-19", Sandra Laugier combines the analysis of vulnerability with the acknowledgement of the importance of women in caring for the most vulnerable during the pandemic (and other major catastrophes). Laugier identifies three areas in which the pandemic might have taught us something: care, global ethics and intersectionality. Giovanni Stanghellini's and Massimiliano Aragona's "Phenomenology of Vulnerability: A Person-Centred Approach" explores the use of the notion in medicine and health, to show how the reduction of the existing categorical and contextual/situational concepts of vulnerability to a quantitative approach falls short of accounting for personal reactions to challenging events. The authors call for a shift of focus towards the ways individuals cope with such contingencies. Two contributions, "(In)visibility of Children and Their Psychosocial Vulnerability—The Narrowed Discourse on Children in the First Year of the Pandemic in Germany", by Maria Griemmert and Anne Oommen-Halbach, and "Social Inequality in Child Health and Development—Before and After the COVID-19 Pandemic" by Simone Weyers and Mariann Rigò, consider the pandemic fate of children. In the first essay, the focus is on the portrayal of children and young adults' condition by German media, during the first year of the pandemic. The emphasis here is on the ambivalent representation of this group in Germany, but the same attitude can be found in most Western countries: they were mostly banished from public discourse and considered risk factors for the more vulnerable populations such as the elderly, while their own psychosocial vulnerability was completely marginalised. Weyers and Rigò turn their attention to children growing up in families with a low socioeconomic position, in which "health inequalities" become especially evident. The authors highlight the consequences of the pandemic on these already vulnerable individuals, further weakened by the closing of essential health-promoting facilities, such as kindergartens and schools. Building on a significant statistical survey on the sharp decrease in school enrolment of precisely the most vulnerable children during and after the pandemic, the authors call for greater preventive efforts for this group. Chiara Montalti's essay, "The (Crip) Art of Reworking Vulnerability—and Perhaps, to Find a Way Out of It", confronts vulnerability with disability. It addresses their connexion by taking collective caring relations as a case study. The epistemological and political perspective of the essay intersects these two concepts in a fruitful way, not flattening them on each other, but stressing their common relational nature, and showing how positive social responses can beneficially affect both experiences. The essay "Total Institutions" as Litmus Test of Civilisation" by Chantal Marazia, Nils Löffelbein and Heiner Fangerau builds on the media coverage of what Erving Goffman has called "total institutions" at the outbreak of the last pandemic. The authors read this conspicuousness and the popularity enjoyed by Goffman's concept itself during and the immediate aftermath of the pandemic as an indicator of a perspective-shift away from vulnerable subjects and towards the contexts that make some people more fragile than others.

This essay, in a sense, foreshadows the last part, opened by Ubaka Ogbogu's "Pandemic Necropolitics: Vulnerability, Resilience, and the Crisis of Marginalisation in the Liberal Democratic State". Drawing on inputs from Achille Mbembe, especially his category of "necropolitics", the paper sheds light on the extent to which marginalisation and vulnerability are the product of a covert and intentional politics of death. By rejecting the idea of resilience, so frequently identified as a way to overcome the difficulties caused by the pandemic, the author regards the overturn of the necropolitical condition as the only possible solution.

In his "Vulnerable to Ourselves, or the Radicalised Disenchantment of Being", Kamran Baradan moves along similar lines. Building on the analyses of Jacques Lacan, Judith Butler and Slavoj Žižek, Baradan highlights the role of vulnerability in the structuring of subjectivity, inter-subjective relations and politics. He also suggests that the "vicious circle of vulnerability" be broken by a new definition of political act, in order to abolish and deconstruct capitalist relations and the very circumstances that make us vulnerable. Henk ten Have's "Vulnerability as a New Perspective on Ethical Challenges in Healthcare" takes a fresh look at the bioethical notion of vulnerability, by scrutinising its use in ethical discourses on health and health care. Then, ten Have uses the pandemic scenario as the key to a deeper reflection on vulnerability as an anthropological, but especially as a politically constructed and enforced condition. It is to this latter dimension that this chapter invites to draw attention, shifting the focus to its structural, socio-political determinants.

The essay by Jane Cooper, Zamina Mithani and Wesley Boyd, "Vulnerability, Interest Convergence, and the COVID-19 Pandemic: Lessons from the Future", closes the volume, reiterating the urge to exploit the pandemic emergency as an ethical laboratory of sorts, with the potential to reveal ways of reimagining bioethics altogether. Combining established descriptive (Rawls' "veil of ignorance"; Bell's "convergence of interests") and novel prescriptive models (Heckler and Mackey's "interest cognizance"), the authors urge for transcending a mere retrospective ethical analysis of the pandemic, towards a culture of ethical preparedness, open to the countless global challenges still facing us.

Not by chance, the title of this Part III is *The Future of Vulnerability*, reflecting what needs to be an active hope to strengthen vulnerability as a hermeneutic, epistemological, critical-normative perspective and the auspice that the future belongs to vulnerability. This means that the care of the vulnerable—in its manifold configurations and meanings—must grow into an essential feature of a society departing from core values rooted into domination, competition and inequality, towards a culture of justice, inclusiveness and respect for differences. In this sense, our reference to "Post-pandemic" in the title must be intended less as a chronological marker (the post-pandemic time proper having just about started), but rather in the programmatic sense of an authentic caesura, one to be not simply observed, or prophesised, but actively sought.

We would like to thank all contributors for having taken care of that portion of vulnerability that always marks an intellectual path in research that intends to authentically question the problems of its time.

Stefania Achella
Chantal Marazia

References

1. Honigsbaum M (2020) The pandemic century: a history of global contagion from the spanish flu to COVID-19. Penguin Random House, London
2. Brown K (2014) Questioning the vulnerability zeitgeist: care and control practices with "vulnerable" young people. Soc Policy Soc 34(3):371–87
3. Pupavac V (2001) Therapeutic governance: psycho-social intervention and trauma risk management. Disasters 25(4):358–72
4. Fineman MA (2010) The vulnerable subject and the responsive state. Emory Law J 60:251–275
5. Burgorgue-Larsen L (2014) La vulnérabilité saisie par la philosophie, la sociologie et le droit de la nécessité d'un dialogue interdisciplinaire. In: Burgorgue-Larsen L (ed) La vulnérabilité saisie par les juges en Europe. Pedone, Paris, pp 237–243
6. Brown K, Ecclestone K, Emmel N (2017) The many faces of vulnerability. Soc Policy Soc 16(3):497–510
7. Polychroniou A (2022) Towards a radical feminist resignification of vulnerability: a critical juxtaposition of Judith Butler's post-structuralist philosophy and Martha Fineman's legal theory. Redescriptions: Polit Thought, Conceptual Hist Feminist Theory 25(2):113–136
8. Cavarero A (2009) Horrorism: naming contemporary violence. Columbia University Press, New York
9. Gilson E (2014) The ethics of vulnerability: a feminist analysis of social life and practice. Routledge, New York
10. Luna F (2009) Elucidating the concept of vulnerability: layers not labels. Int J Feminist Approaches Bioeth 2(1):121–39
11. Wrigley A (2015) An eliminativist approach to vulnerability. Bioethics 29(7):478–489
12. Flood CM, MacDonnell V, Philpott J, Thériault S, Venkatapuram S (eds) (2020) Vulnerable. The law, policy and ethics of COVID-19. University of Ottawa Press, Ottawa
13. OECD (2020) Building back better: a sustainable, resilient recovery after COVID-19. Available at: https://read.oecd-ilibrary.org/view/?ref=133_133639-s08q2ridhf&title=Building-back-better-_A-sustainable-resilient-recovery-after-Covid-19&_ga=2.104523456.1389674155.1632825099-113007773.1632825099
14. Schwab K, Malleret T (2020) COVID-19: the great reset. Amazon Digital Services LLC—KDP Print
15. Dennis MJ, Ishmaev G, Umbrello S, Van den Hoven J (2022) Values for a post-pandemic future. Springer, Cham

Contents

Editors and Contributors

About the Editors

Stefania Achella is Professor of History of Philosophy at the University of Chieti-Pescara (Italy), where she teaches History of Philosophy. She is currently working on the relationship between philosophy and biology in the nineteenth century, on the concepts of life and corporeality in classical German philosophy, and on the contributions of Feminism to the philosophical debate. Among her most recent publications: "Idealism and Science of Life: An Intersection Between Philosophy and Biology" (Springer 2022); "Limit Situations: For an existential anthropology of vulnerability" (Rivista di Studi Jaspersiani 2022); *The Owl's Flight. Hegel's Legacy to Contemporary Philosophy* (co-editor, De Gruyter 2021); "Gendering the Anthropocene?" (Itinerari, 2020).

Chantal Marazia is assistant professor in the History Philosophy and Ethics of Medicine at the University of Düsseldorf (Germany). Her research focuses are the history of psychiatry and neurology, and the history of human rights. She is editor of the psychopathological writings of Karl Jaspers.

Contributors

Stefania Achella University of Chieti-Pescara, Chieti, Italy

Massimiliano Aragona The Italian National Institute for Health, Migration and Poverty (INMP), Rome, Italy;
The Roman Circle of Psychopathology, Rome, Italy

Kamran Baradaran Department of Teaching English as a Foreign Language, Faculty of Foreign Languages, North Tehran Branch, Islamic Azad University, Tehran, Iran

Jane Cooper University of Toronto Faculty of Law, Toronto, Canada

Piergiorgio Donatelli Sapienza Università di Roma, Rome, Italy

Anna Donise University of Naples "Federico II", Naples, Italy

Nico Dragano Institute of Medical Sociology, Centre for Health and Society, Medical Faculty and University Hospital, University of Düsseldorf, Düsseldorf, Germany

Heiner Fangerau Department of the History, Philosophy and Ethics of Medicine, Medical Faculty, Heinrich-Heine-Universität Düsseldorf, Düsseldorf, Germany

Rita Fulco Dipartimento di Civiltà Antiche e Moderne (DiCAM), Università degli Studi di Messina, Messina, Italy

Maria Griemmert Department of the History, Philosophy and Ethics of Medicine, Medical Faculty, Heinrich-Heine-University Düsseldorf, Düsseldorf, Germany

Henk ten Have Center for Global Health Ethics, Duquesne University, Pittsburgh, USA;
Department of Bioethics, Anahuac University Mexico, Mexico City, Mexico

Sandra Laugier Institut Universitaire de France, Institut des sciences juridique et philosophique de la Sorbonne UMR 8103, CNRS Paris 1, Université Paris 1 Panthéon Sorbonne, Paris, France

Nils Löffelbein Department of the History, Philosophy and Ethics of Medicine, Medical Faculty, Heinrich-Heine-Universität Düsseldorf, Düsseldorf, Germany

Chantal Marazia Department of the History, Philosophy and Ethics of Medicine, Medical Faculty, Heinrich-Heine-Universität Düsseldorf, Düsseldorf, Germany

Zamina Mithani University of British Columbia Faculty of Medicine, Vancouver, Canada

Chiara Montalti University of Salento, Lecce, Italy

Ubaka Ogbogu Faculty of Law, University of Alberta, Edmonton, Canada

Anne Oommen-Halbach Department of the History, Philosophy and Ethics of Medicine, Medical Faculty, Heinrich-Heine-University Düsseldorf, Düsseldorf, Germany

Mariann Rigó Institute of Medical Sociology, Centre for Health and Society, Medical Faculty and University Hospital Düsseldorf, Heinrich-Heine-University Düsseldorf, Düsseldorf, Germany

Giovanni Stanghellini Department of Health Sciences, University of Florence, Florence, Italy;
"Diego Portales" University, Santiago, Chile

J. Wesley Boyd Baylor College of Medicine, Houston, USA;
Department of Global Health and Social Medicine, Harvard Medical School, Boston, USA

Simone Weyers Institute of Medical Sociology, Centre for Health and Society, Medical Faculty and University Hospital Düsseldorf, Heinrich-Heine-University Düsseldorf, Düsseldorf, Germany

Part I
What Is Vulnerability?

Chapter 1
Vulnerability is Said in Many Ways

Stefania Achella

> *As the other of violence, vulnerability may itself*
> *constitute or be constituted by violence*
> *Eleine P. Miller, Bodies and the Power of Vulnerability, 2002*

Abstract The pandemic has demonstrated, in an amplified way, that vulnerability is common to all human beings. Nevertheless, the paradigm shift proposed by the ethics and discourses built around vulnerability did not take hold. The call for a more just and less competitive society, which became widespread during the pandemic, did not immediately seem to be as urgent afterwards. In fact, since the pandemic, attitudes seem to reflect a desire to erase this experience and restore the *status quo ante*. Can this operational limit of the concept of vulnerability be a sufficient reason to abandon this interpretive paradigm and return to the old Promethean idea of man which shaped modern society? In the light of the pandemic, is it not worthwhile instead to question vulnerability more deeply? Is it not important to continue reflecting on the anthropological approach underlying vulnerability in order to understand how it can provide an ethical and political perspective for building a better future? Starting from these questions, this essay explores vulnerability from a philosophical perspective, analysing its strengths and weaknesses.

Keywords Vulnerable turn · Normativity · Critical theory · Forms of life

1 Introduction

Human beings have been confronting their own vulnerability since ancient times. A vivid embodiment of it in the Western imagination is Achilles, the invincible, generous warrior, without uncertainties, but also the hapless hero, secretly vulnerable. He moves through the world literally walking on his own weaknesses and is

S. Achella (✉)
University of Chieti-Pescara, Chieti, Italy
e-mail: stefania.achella@unich.it

S. Achella and C. Marazia (eds.), *Vulnerabilities*, Integrated Science 18,
https://doi.org/10.1007/978-3-031-39378-5_1

Fig. 1 Iloupersis (the fall of Troy), detail. Side A from an Attic red-figure kylix, ca. 490 BC. From Vulci (Louvre Museum), photo of Marie-Lan Nguyen (public domain)

finally killed, treacherously, precisely because of that heel, which deeply marked his humanity and therefore his vulnerability (Fig. 1).

Achilles is the symbol of that humanity in relation to which he was stronger, but to which, despite his mother's efforts, he could never cease to belong.

Even more peculiar is the figure of Antaeus. The son of Gea and Poseidon, like the other Giants he had extraordinary strength, although unlike his brothers, this strength had the characteristic of increasing the moment he fell to the ground wounded, and re-established contact with his mother, Gea. Antaeus represents both vulnerability and resilience. After being injured, the giant literally gets up again—in the same sense we find in the Latin expression *resilire*.[1] Each time he falls, Antaeus becomes stronger: his strength is nourished precisely by his weakness. As these two examples show us, in contrast to the later tradition, the male who embodies fragility in the Greek tradition often encounters an attitude of acceptance or even an embracing of vulnerability in order to achieve virtue. In classical Greece, vulnerability refers not only to the possibility of being wounded but also to the "self-awareness and acceptance of being subject to harm" [2, p. 205]. Moreover, the staging of this vulnerability in a tragedy, an epic or in philosophy invites the political community to witness fragility and thus to include the vulnerable more fully.

It is no different in religions. Think of Christianity, where even God experiences vulnerability by incarnating himself in human form and suffering the wounds of crucifixion; or Buddhism, where the human experience of physical vulnerability becomes a path to wisdom—as the story of the young prince Siddhartha and his decision to embark on the path to enlightenment after experiencing human suffering (*Duḥkha*) in the form of ageing, illness and death attests.[2]

---

[1] The Latin expression is formed by adding the prefix *re* to the verb *salire* 'to leap, to bound, to bounce', and thus means 'to spring back, to rebound'. On the bound of resilience and vulnerability [1].

[2] As Zhang states: "The word 'suffering' is a translation of the word duhkha (Pali) or dukkha (Sanskrit), which literally means dis-ease or unsatisfactoriness. There is a well-known Buddhist

The attitude in modern Western culture has changed. It tries to remove from human nature the humble aspect, everything that relates to deficiency, exposure, lack, disease, and even death. Those who embody vulnerability become the object of passions such as disgust, contempt, and even fear. It was the English philosopher Thomas Hobbes who inaugurated a long tradition in the political field. He showed that political theory is uncomfortable with vulnerability, and indeed does everything it can to expel it from this domain. The pact of delegation to the sovereign, of submission to the politician, is made in the name of the extradition from everyday life of every form of vulnerability to which living with others exposes us (violence, traps, ambushes, etc.). Hereafter, when politics confronts vulnerability, "it is to virilely place it at a distance, somewhere 'prior to' or 'alongside' the political game, or to bury it among its silent drudgery by hastening to label it with names less stamped with powerlessness" [3, p. 7].

Neither does contemporary politics seem to escape this discourse in which power is maintained by containing or eradicating vulnerability through the biological metaphor of immunisation. This term, introduced into philosophical-political language by the Italian philosopher Roberto Esposito, indicates the process of increasing defence of the community against infiltration and contagion of foreign elements.[3]

Borrowing an expression taken from the Lacanian lexicon, Judith Butler also denounced the ethical implications of Western secular culture as a process of 'foreclosure' of the elements of vulnerability. In other words, it is the exclusion which the subject chooses to put in place in order to gain an acceptable, intelligible identity that is worthy of recognition. It is not simply a mechanism of repression of something that would remain 'inside', or the removal of something that is destined to return, but rather a process of structural exclusion which is simultaneously vital, yet also pathogenic, for the subject.

The *heuristics of vulnerability*, developed at the end of the last century, therefore opposes the pale image of a subject that would only be built on strength and security and seeks instead to bring to the fore the ontological, and as such ineradicable and universal dimension of vulnerability. Rather than nurturing the essentialisation of an intact, healthy subject, it asserts the need to promote the interrelational aspect, the

---

claim, "All this is dukkha." Suffering is, then, shown as a kind of dis-ease caused by human finitude. However, suffering is more complicated than a subjective, psychological description or an intentionalist view that the phenomenal character of any experience is entirely constituted by its representational content; instead, it has a wide range of meaning from that experienced and reality itself, although Buddhism does not seem to focus on reality as it is without human experience" [46, p. 43].

[3] This aspect is further developed in the light of the recent pandemic. In his most recent books, written during and after the pandemic, the philosopher tries to imagine a different kind of immunisation, which seems to lose its constrictive connotations and requires a new interpretation, both biological and political. He shows different immunising reactions which can take different account of vulnerabilities. For example, the model of *herd immunity* proposed at the beginning of the pandemic by the United Kingdom, Sweden, the United States and Brazil is based on tanatopolitical principles that envisage, if not the elimination, at least the marginalisation of the "less fit" in favour of the more productive segments of the population [see 47].

mutual dependence by virtue of the vulnerability into which we are all already thrown. The 'vulnerable subject' seems to offer a powerful alternative to the mythical liberal autonomous subject of neo-liberal rhetoric [4, p. 504]. To the dominant political and legal subject in the modern age, which describes the human being as a competent, capable, self-sufficient and self-fulfilling agent "who seeks liberty or autonomy as a primary value" [5, p. 108], a relational subjectivity was opposed. In contrast to liberal constructions of the subject, this [6, 7] *heuristic* has insisted on the embodied nature of the vulnerable subject and it is a priori belonging to sociomaterial contexts. These contexts not only produce vulnerability but also determine its degree of resilience: recognising vulnerability means both setting in motion precisely different qualities of relationship and a more inclusive, embodied and interactive social model [see 6, 7].

The pervasiveness of vulnerability as a heuristic model can also be seen in the numerous official acts of major supranational organisations[4] which have sought to reorient institutional interventions in order to concretise the commitment to protect individuals—in their autonomy, dignity and integrity—from threats to personal fulfillment. By promoting a model of citizenship based on interdependence, empathy and the foregrounding of social-ethical obligations to others [see 8], vulnerability has seemed able to circumvent many of the shortcomings of previous efforts to show the political roots of harm and suffering [see 9]. The pandemic provided irrefutable evidence of the reasons for the ethics and heuristics of vulnerability, namely, the impossibility of overcoming our fragile condition alone. Many studies comparing the reactions in the months of the pandemic and in the first months after the emergency have confirmed this perception and manifested the need for a change in values in a postmaterialist direction. For example, Lampert's analysis showed that:

> the pandemic and the economic crisis it brought have led to an increased focus on individual free choice and the non-material aspects of life. At the same time, the support for [...] law and order have decreased. People are increasingly calling for inclusive growth and for reducing the gap between rich and the poor. [10, p. 3][5]

However, the signals coming from post-pandemic society seem to indicate that it does not intend to move in that direction. The shift from *homo oeconomicus* to *homo*

---

[4] See: UNESCO's "The Principle of Respect for Human Vulnerability and Personal Integrity," Report of Keep the hyphen International Bioethics Committee of UNESCO (IBC), (2013); the University of South Carolina's Hazards and Vulnerability Research Institute "Social Vulnerability Index for the United States" (2013); the Council for International Organisations of Medical Sciences' "International Ethical Guidelines for Biomedical Research Involving Human Subjects," prepared in collaboration with the World Health Organisation, (2002); and going further back, the "Barcelona Declaration on Policy Proposals to the European Commission on Basic Ethical Principles in Bioethics and Biolaw", adopted in November 1998, and the NIH's "The Belmont Report: Ethical Principles and Guidelines for the Protection of Human Subjects of Research" (1979).

[5] Ibo van de Poel, Tristan de Wildt, and Dyami van Kooten Pássaro, in their study [48, p. 47] used the computational tool of topic modelling - which allows one to track the changing frequency of specific topics in a corpus of text. The results showed that while the values of safety and health increased significantly in the first months of the pandemic, the values of democracy, privacy and socio-economic equality decreased.

vulnerabilis, which seemed inexorable until a few months ago, no longer seems to be on the agenda. What has gone wrong, why has this trend come to a halt? Let us consider the main questions raised by the discourse on vulnerability.

2 Different Meanings of Vulnerability

In their important study, Catriona Mackenzie, Susan Dodds, and Wendy Rogers identified three types of vulnerability: *inherent*, *situational* and *pathogenetic* [11, p. 24]. The first type has to do with our bodies, [13][6] the second with the historical, geographical and environmental context, and finally the third with the political and social politics that can create injustice.[7] Just as the body has pressing material needs that expose us to illness, disability and death, so as social and affective beings, we experience loss, bereavement, abuse, lack of care, rejection and humiliation. Politically, we are vulnerable because we are subject to exploitation and manipulation, rights violations and political violence. All of these forms of vulnerability are manifested to a greater or lesser extent depending on what Castel called "supports".[8] Regardless of the type of vulnerability that may affect us, we are all subject to a "social property" that includes rights, resources and protections that cannot be disregarded [see 13].[9]

There are different perspectives from which to view vulnerability, emphasising the biological or psychological aspect or the ethical, social or political aspect. For the purposes of our analysis, however, we shall distinguish between two categories: on the one hand, that which interprets vulnerability in its *broadest* sense. It concerns human beings in general and is linked to the biological fragility of the body, the existential perception of one's own finiteness [16, 17] and the constant exposure to others, both emotionally and materially (Fig. 2).

[6] This may include what Fineman calls "the human being's embodied vulnerability", which varies according to the quality and quantity of resources we possess or can use in order to also be resilient to those elements that make us vulnerable [37, p. 21].

[7] To quote Catherine Malabou's expression, "Ontology of the Accident" [12].

[8] "Crucially, notes Clough, this is not to say that this shared vulnerability is experienced in the same way. The importance of focusing on the particular experience is a vital aspect of vulnerability theory and recognises, perhaps more clearly than the social model, that it is the particular individual's interaction with society which is significant. This raises further questions of how we can make law and policy responsive to particular individuals and how interventions or shifts in broader structures or institutions would impact on users of services" [49, p. 479].

[9] It should be noted, however, that the view that the risks of the virus could affect anyone, regardless of their location, has been widely criticised as the pandemic is exacerbating existing health inequalities [14]. Furthermore, intersectional studies have shown that when the workforce is racialised and feminised, safety standards decline along with wages. [15]. Even if, as Sandra Laugier notes, the pandemic has "highlighted the vulnerability of everyone, including the privileged, who have found themselves lost without their many 'services' [...] the better-off have the capacity to conceal or deny their acuteness by delegating care" [50, p. 52].

Fig. 2 Salvator Rosa, *Humana Fragilitas*, 1656 Fitzwilliam Museum, Cambridge (particular)

On the other hand, we have a *narrower* meaning that concerns certain individuals or persons who are particularly disadvantaged or fragile under certain conditions, which Anderson defines as "surplus vulnerability" [18, pp. 155–156]. In the first case, vulnerability is synonymous with humanity; it is not a contingent susceptibility that affects specific individuals or categories of people. It is about humanity: as Erving Goffman suggests, it is the whole of ordinary experience that is structurally vulnerable [19]. It has therefore been referred to as "universal vulnerability" or the "vulnerability thesis" [16, p. 1]. In the second case, vulnerability is materially relational: one is vulnerable to particular agents with respect to particular kinds of threats [20, p. 112]. This form of vulnerability is opposed to the idea of autonomy: the more vulnerable a person is, the less autonomy he or she has in terms of being able to realise his or her life goals without necessarily depending on others (people, institutions, tools, etc.). Affiliations are also crucial here, as they can be either a corrosive disadvantage or a substantial help in overcoming vulnerability. The first meaning makes the concept too broad and therefore, according to some, not always useful for ethical normativity. Nonetheless, it helps us to overcome what has been defined as a top-down approach

[21] which would consider vulnerability as a derivative aspect, a reflection of the failure to realise a predetermined positive model. According to the top-down model, one would first define what makes a life good, just and satisfying in order to derive the values that serve to promote its development and ability to flourish. In this case, vulnerability is defined *ex negativo* or *per derivationem*. This last perspective intends vulnerability as a lack, a deficiency of those elements that are fundamental to the attainment of a full life. Such a full life is the point that ethics seeks to reach when it uses the concept of vulnerability in a narrow sense [22]. The *broadest* sense of vulnerability can help us to understand vulnerability not as a problem to be solved but as an inescapable aspect in the definition of life. In doing so, it becomes an opportunity to reassess our view of ourselves and the world rather than a negative obstacle to be overcome or managed. The constant exposure of our lives to rupture, floundering and lack of expression can actually lead to the revelation of new realms of possibility. "In this perspective," as Piergiorgio Donatelli says, "human life is not described through an exhaustive list of criteria that are available beforehand and can be fixed for determinate goals. Human life is rather what we find after a loss, a crisis, a condition of discomfort or uneasiness" [23, p. 1033].

If understood as a defining characteristic of the human being, vulnerability seems to be linked not only to our biological or socio-political weaknesses but also to the inexorable destiny that confronts us with failures, with the impossibility of realising all our perspectives and projects in the course of our lives, but which also allows us to develop active responses to these experiences [24, p. 19].

3 Some Aporias

These two conceptions of vulnerability have had an important *explanatory* function, providing a sufficiently broad framework for the perception of a range of phenomena concerning fields of experience and forms of criticality. However, they have also had a *propositional-constructive*, dynamic function in terms of institutional arrangements, bringing to the fore certain needs that had been denied for centuries. As Estelle Ferrarese has rightly pointed out, if it is true that vulnerability is a perceptible fact, it is equally true that it only comes to light when one is willing to acknowledge it and to act in some way to deal with it. From that moment on, it promotes ethical obligations. Vulnerability, in fact,

> *only appears insofar as it entails a horizon of obligations* (fulfilled or not, but perceived by some and, in any case, by whoever uses the terms) *and of normative reasoning*. It may be a matter of obligations that you attribute to yourself, or that you impute to others [...]. In the latter variation of imputing it to others, the notion of vulnerability may carry with it an injunction for the State to act, an imperative addressed to institutions to protect or palliate. [3, p. 25]

However, by sharpening the focus of the meanings both of vulnerability in a narrow sense (highlighting the plight of the so-called 'vulnerable') and the idea that we are all vulnerable, fragile and powerless in the face of certain life events [25], the pandemic has shown with great clarity the limits and pitfalls of both the *heuristic* and the *normative* uses of this concept. The pitfalls relate to the inability of this appeal to vulnerability to ensure an adequate response or to exclude violence from that response.

Now let us take a closer look at the aporias involved in the use of this concept. The first concerns a project that we could define more broadly as *cultural* and *anthropological*: the development of a culture of fear and of a subject that is considered anthropologically incapable of responding to offence of any kind. The second, which concerns the *sociopolitical* level, refers to the stigmatisation of the most vulnerable categories. To be labelled 'vulnerable' is not to be on an equal footing with other 'nonvulnerable' people. A further aspect could be called *epistemic*, namely, the presence of an epistemological opacity underlying the identification of vulnerable groups to the detriment of others—on this point, think about the difference in status between an economic migrant and a refugee.

The question of the *anthropological* perspective is interesting because it contains an element of ambiguity. In a 2004 paper, Furedi argued that society's emphasis on the recognition of human vulnerability leads to a sense of powerlessness and diminished responsibility, which corresponds to a decrease in autonomy. As Furedi wrote, "contemporary culture answers these questions [about illness and trauma] by stating that everyone needs help and everyone needs support. That is why ideals of independence and self-sufficiency have given way to a culturally sanctioned state of dependency" [26, p. 103]. He continued, "If we renounce the possibility of having some choice over the direction of our life then we risk diminishing the meaning of our humanity" [26, p. 194].[10] Alain Badiou's criticism was along the same lines. He saw in the contemporary discourse on the defence of human rights a process of victimisation that reduces human beings to victim animals rather than active subjects. In other words, the rhetoric of vulnerability has led to thinking about human beings in terms of suffering and death, with the result that all constructive potential is lost [28, pp. 10–13]. According to this interpretation, in the gradual shift from a dimension that emphasised elements such as resilience and rationality to a narrative that emphasises vulnerability instead, there has been a reconfiguration of existential, social and political expectations: "The widespread acceptance of a vulnerability model of the

---

[10] As one can read in another text of Furedi of 2003: "The model of human vulnerability and powerlessness transmitted through therapeutics coincides with a far wider tendency to dismiss the potential for people exercising control over their lives. The narrative of emotional vulnerability coexists with powerful ideas that call into question people's capacity to assume a measure of control over their affairs. Social commentators regularly declare that we live in the era of the 'death of the subject', 'the death of the author' or the decline of agency. Such pessimistic accounts of the human potential inform both intellectual and cultural life in the west. The survivalist outlook alluded to by Lasch is not simply fueled by a preoccupation with the vulnerability of the self but also by the conviction that the world has become an intensely dangerous place beyond the control of humanity. Western society is continually haunted by the expectation of crisis and catastrophe. Environmental disasters, weapons of mass destruction, 'technology gone mad' are just some of the concerns that have helped to fashion a permanent sense of crisis" [51, p. 130]. See also [52, p. 57]. More recently, also [27]. This discourse was also at the centre of Giorgio Agamben's reflections on the proposed restrictions during the COVID 19 pandemic [see 53].

human being and a focus on victims are a corollary of the decline of beliefs about the perfectibility of society and the rise of an ethos of 'no alternative'" [29, p. 11]. This analysis therefore highlights the risk of using the universalist conception of vulnerability:

> By emphasising its universality and amplifying its generative capacity [...the idea of a universal vulnerability, SA] might unwittingly dilute perceptions of inequality and muddle important distinctions among particular vulnerabilities, as well as differences between those who are injurable and those who are already injured. [9, p. 262]

In response to these concerns, we can evoke what was underlined by the feminist tradition. It has encouraged the abandonment of the concept of autonomy as a central value in ethical discourse. In contrast to the ideal of autonomy that animates most contemporary moral theories, Marlène Jouan and Sandra Laugier remind us that we need others as well as ourselves to satisfy our basic needs and that dependence is therefore an essential element of the human condition [30].

Other aporias are related to the insistence on the specific vulnerability of certain categories of subjects. If, in fact, the identification of a specific vulnerability is important in order to provide more adequate assistance to people in difficulty, such identification has contributed to fuelling new and dangerous categorisations by emphasising the exercise of social control and paternalistic intervention in the lives of those classified as vulnerable. Moreover, pathogenic forms of vulnerability can also be the result of a response initially designed to resolve a vulnerability, which instead paradoxically exacerbates it or produces new forms of vulnerability [11, p. 9], leading to the idea of an erosion of the role of collective movements and an expansion of social control by the state [cf. 31–33]:

> Characterising a population as 'vulnerable' can have the effect of stigmatising it, thereby justifying forms of segregation, discrimination or tutelage. In this way, the scientific gesture amounts to constructing a paradigmatic subject and endeavouring to identify groups that do not correspond to this paradigm as vulnerable; to the extent that they are frequently declared such owing to a compromised or dubious capacity to consent, it is easy to make out an effect of subjugation. [3, p. 16]

Finally, there is the epistemological question. At the end of the last century, Judith Butler pointed out that the exposure of the concept of vulnerability to the normative contexts in which it develops could represent another problematic element. It is no coincidence that the American philosopher drew attention to the difference between vulnerability and mourning for the lives destroyed in the bombing of the Twin Towers and for the millions killed in the war in Afghanistan. Butler invited us to pay attention to the epistemic framework, cultural in the strong sense, understood as the structuring totality within which the dynamics unfold that lead to the identification of subjects as 'vulnerable' and therefore worthy of help, support and mourning. This is an idea of vulnerability the semantic sphere of which seems to be related to the concepts of authenticity and/or individual integrity. This use of the concept of vulnerability presupposes an idea of the subject that is placed within pre-existing frameworks of intelligibility that decide on the visible and the invisible, the sayable and the unsayable, the representable and the unrepresentable, on what is worthy and what is not worthy of being recognised, both in its qualities and in its fragilities. From this perspective, it is clear that vulnerability is placed inside or outside the processes of recognition, processes that will become increasingly evident, integrating in more or less time the demands and needs of some subjects and negating those of others. Here, the risk highlighted by Butler is that of a re-essentialisation of vulnerability in terms of a homogeneous and unified identity. In this regard, Butler critically argues that the institutional recognition of a social subject's vulnerability depends on the hegemonic frames of recognition and dominant patterns of intelligibility that authoritatively define, in every sociocultural contextualisation, who counts as human and who does not.

4 Ethical Models of Vulnerability

How can these aporias be overcome? Recently developed ethical models have attempted to overcome some of these limitations [for this distinction, see 3].

The first model, theorised by Butler [34] and Cavarero [35], consists of proposing an ethics based on noninjury. That is, an *ethics of nonviolence* is derived from the anthropology of vulnerability. Such an ethics develops a different meaning of responsibility, also nourished by the common condition of vulnerability, from which a minimum imperative of solidarity should be developed. Starting from the suffering that our bodies endure at the behest of another body, which leads us to an experience of pain—understood as that which makes us aware of our interdependence and inescapable shared vulnerability—Butler configures an ontology of vulnerability in opposition to the individualist ontology of modernity and to the individualist claim to a self that is untouchable by the other. Our exposure to the other, to outrage and violence, allows us to respond to the other who challenges us and invites us to take responsibility. Responsibility, in fact, does not depend on will but is the result of the inevitable vulnerability that allows us to respond to the other. Violence reveals our structural physical vulnerability in a private and public exposure to the action of the other from which we cannot escape and which makes us aware that we are not isolated individuals. This model seems to answer the questions raised by Furedi about the idea of a weakened and fearful humanity. In this case, vulnerability is understood as a sense of strength. The often-cited limitations of this approach are that it reduces vulnerability to an ethical and non-political issue (see [4]).

The second model is related to the ethics of *care*. It is not about refraining from harming the other but about the active, positive duty to perform an act of care towards the other. Care, as Joan Tronto [36] has well pointed out, refers to a whole field of social interventions aimed at alleviating the suffering of the vulnerable and ensuring the best possible quality of life. However, according to Tronto, it also includes all kinds of everyday gestures that contribute to one's own well-being and that of the community in which one lives. This attitude challenges the classical doctrines of moral philosophy, which largely ignore this type of practice or at least subsume it under a set of theoretical concerns that they consider higher, relegating it to feminist thought and women's practice. In contrast, we can say that the long process of maturation and evolution, combined with the properly human capacities of moral perception and attachment, make caring for people who are dependent on others the mark of our humanity. Caring, then, concerns not only interpersonal relations and social justice but also the level of political intervention: caring, giving and receiving care complete the circle that unites the individual and the collective, at the centre of which is the subject as vulnerable and potentially dependent, i.e., all human beings. One of the limitations of this perspective has been to consider care as primarily a moral paradigm.

Finally, there is a third model. It is related to the interpretation of vulnerability as an *impropriety of the self*. Fineman's model [37], which is linked to Nussbaum's idea of vulnerability [38] related to fate, accidents and violence, considers the intervention of institutions necessary but places the greatest responsibility on the vulnerable subject, who has the obligation to react (to be resilient), thus favouring social adaptation over criticism or transformation of existing structures and social relations. These models, which give a good account of the constant intertwining of the psychological and moral levels with the bodily level, encounter a limitation: 'a distancing of the political, either through its forgetting or its strict limitation' [3, p. 38] or, to use Shulman's words, "they do not ask how injury can be transformed into action" [39, p. 235].

The limit of these proposals, as Ferrarese [3] has pointed out, would therefore be the marginal role of the political dimension, which leaves the solution to vulnerability to the individual or to interpersonal relationships. Is it possible to identify a different trajectory to at least partially overcome some of the limitations of this proposal?

5 Vulnerability as a "Critic to Forms of Life"?

One way in which critical theory [3, 41] responds to these limitations is by proposing a critical analysis of vulnerability.

Suffering was already used in this sense by the early Frankfurt School theorists: from its epistemological status, suffering derives its political status because it pushes for social transformation: "The physical moment tells our knowledge that suffering should not be, that things should be different" [40, p. 203; on the capacity of suffering to be a form of emancipation, see 41].

Perhaps a further step in this direction can be taken by combining the concept of vulnerability with that of forms of life and, more specifically, by considering vulnerability as a *critique of forms of life* [see 42].

As mentioned above, the use of vulnerability as a critical and normative category has been extensively developed by the critical theory approach. Similarly, the concept of "form of life", or rather of "forms of life" (used in the plural because it does not only refer to the biological dimension), has also found considerable elaboration within the more recent developments of this tradition (Fig. 3).

Forms of life have been understood as the way in which an individual or a group of individuals live their lives, either by virtue of their biological constitution or on the

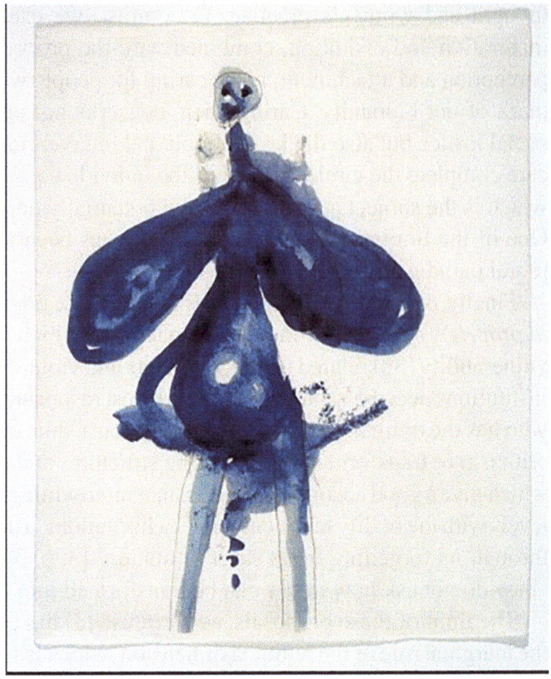

Fig. 3 Louise Bourgeois, *The fragile*. *Source* Wikiart

basis of values, beliefs, and habits in which the subjects are already located (hence Butler's idea of subjectivation and domination). It may be useful here to return to the notion of forms of life as it has been reconsidered in recent thinking, insofar as it overcomes the distinction between the social and the biological dimension and allows greater attention to be paid precisely to their articulation. The forms of life of an individual or group are derived from their biological constitutions, but those forms also shape their experiences, thoughts and actions, which in turn are shaped by the culture and society in which they live. It has to do with the ethical and historical dimension, but it also shows the circular relationship, as Hannah Arendt pointed out, between the world and human life, which mutually shape each other. Finally, forms of life are also attempts to solve problems, crises and conflicts. However, they are not ways of life (*Lebenweisen*), they are not individual options, rather they are about both the context from which we come and that in which we are formed. It is precisely because of the comprehensiveness that the concept of forms of life, when combined with vulnerability, can offer a resource which is not only hermeneutic and normative but which is also a source of emancipation. Vulnerability thus becomes a way of showing what is wrong with our forms of life: existential crises, project failures, alienation but also, physical, economic, cultural hardships, the exercise of power, domination, etc.[11] In this sense, vulnerability becomes a critical category that

[11] In her work on forms of life as a critique of capitalism, Rahel Jaeggi underlines the primacy given to critical activity per se, which in turn is focused on crises and problems, and thus the relaunching

allows tensions and ambiguities to emerge, as well as the richness of relationships. On the other hand, the concept of vulnerability can also outline the positive tasks by which forms of life are measured, the success of which is understood precisely in terms of resolving crises and conflicts, thus avoiding any essentialism, paternalism or perfectionism.

Vulnerability as a critique of forms of life is thus revealed as a search for the conditions of possibility for the transformation and appropriation of conditions of life. Considering that forms of life are "materialised" in institutions, and even more in architectures, tools, bodies and material structures, we understand how they set the limits of what we can do while allowing us to do things in a certain way. Thus, if forms of life refer both to a constitution of their own and to the product of the elaboration of what happens in society, vulnerability as a form of critique can help us to work towards a structural transformation of individual and social practices and institutions. The novelty that it introduces in relation to previous forms of criticism is that it refers to an idea of the human being—as vulnerable—that is completely different from that one that has been dominant in the Western tradition and that has also permeated part of critical theory.

In this sense, vulnerability can respond to attempts at essentialisation by positioning itself as a non-essential, non-exclusive and non-constraining condition of lived, plurally stratified existence [44]. That is, even if the transition to *homo vulnerabilis* does not take place, the ability to look at the vulnerabilities [45] of all our forms of life from a critical perspective could still serve to denounce and act on multiple levels and in a differentiated way with respect to each individual vulnerability. In fact, to understand vulnerability as a critique of forms of life does not mean to give it a precise sphere of intervention, or of action, or a rigid form, but to make it an open, unconditioned and unconditional tool of analysis and of policy and practice. *Vulnerability as a critique of forms of life* can thus assume the ability to make the normative expectations associated with it explicit, to show new noncodified forms of vulnerability, to make them reflexive, and to make them the subject of debate. Additionally, it can be a way of representing the reasons for conflicts and struggles in society, thus fully recovering the claim of the new subjectivities in the public sphere to be recognised.

Core Messages

- The concept of human vulnerability has been revived in recent decades to overturn the classical anthropological vision of the Promethean human being.
- The revival of this concept in many fields and with many variations has made it too broad, depriving it of its effectiveness.
- Both the heuristic and the normative use of the concept of vulnerability have thus far proved inadequate.

---

of a "negativist" approach with regard to all those philosophical and anthropological attitudes that instead aim at researching and identifying the hypothetical essential or fundamental nuclei of human existence or its (self-)realisation. [43]

- Vulnerability as a critique of forms of life can help to build a socio-political model that is more respectful of justice and equality but also of the uniqueness and needs of each living being.

References

1. Miller F, Osbahr H, Boyd E, Thomalla F, Bharwani S, Ziervogel G, Walker B et al (2010) Resilience and vulnerability: complementary or conflicting concepts? Ecol Soc 15(3). http://www.jstor.org/stable/26268184.
2. Berzins McCoy M (2013) Wounded heroes: vulnerability as a virtue in Ancient Greek literature and philosophy. Oxford University Press, Oxford
3. Ferrarese E (2016) Vulnerability and critical theory. Crit Theo 1(2):1–88
4. Brown K, Ecclestone K, Emmel N (2017) The many faces of vulnerability. Soc Policy Soc 16(3):497–510
5. Fineman MA (2015) Equality and difference—the restrained state. Emory Legal Stud Res Paper 15:100–117
6. Wishart G (2003) The sexual abuse of people with learning difficulties: do we need a social model approach to vulnerability? J Adult Protect 5(3):14–27
7. Clough B (2015) Exploring the potential of relational approaches to mental capacity law, https://www.escholar.manchester.ac.uk/uk-ac-man-scw:266067 (visited 03.01.2022)
8. Goodley D (2014) Dis/ability studies: theorising disablism and ableism, 1st edn. Routledge, London
9. Cole A (2016) All of us are vulnerable, but some are more vulnerable than others: the political ambiguity of vulnerability studies, an ambivalent critique. Critical Horizons 17(2):260–277
10. Lampert M, Inglehart R, Metaal S, Schoemaker H & Papadongonas P (2021) Two faces of Covid-19 impact: the pandemic ignites fear, but boosts progressive ideals and calls for inclusive economic growth. Measuring the pandemic's impact on social values, emotions and priorities in 24 countries. Retrieved from https://glocalities.com/latest/reports/valuestrends: Glocalities
11. Mackenzie C, Rogers G, Dodds S (2014) Vulnerability new essays in ethics and feminist philosophy. Oxford University Press, Oxford
12. Malabou C (2013) Ontology of the accident: an essay on destructive plasticity, trans. Shread C. Polity Press, Cambridge, UK
13. Castel R (1991) 'De l'indigence à l'exclusion, la désaffiliation', Précarité du travail et vulnerabilité relationnelle. In: Donzelot J (ed) Face à l'exclusion: le modèle français. Esprit, Paris, pp 167–168
14. Main S (2020) COVID-19 and health disparities: the reality of 'the great equalizer.' J Gen Intern Med 14:1–2
15. Das Gupta T (2006) Racism/anti-racism, precarious employment and unions. In: Vosko LF (ed) Precarious employment: understanding labour market insecurity in Canada. McGill-Queen's University Press, Montreal, pp 333–349
16. Fineman MA (2008) The vulnerable subject: anchoring equality in the human condition. Yale J Law Fem 20(1):1–23
17. Gilson E (2014) The ethics of vulnerability: a feminist analysis of social life and practice. Routledge, New York
18. Anderson J (2014) Autonomy and vulnerability entwined. In [16], pp 134–161
19. Goffman E (1981) Form of talk. University of Pennsylvania Press, Pennsylvania
20. Goodin RE (1985) Protecting the vulnerable: a reanalysis of our social responsibilities. University of Chicago Press, Chicago
21. Donatelli P (2021) The politics of human life: rethinking subjectivity. Routledge, New York
22. Rogers W (2014) Vulnerability and Bioethics. In [16], pp 60–87

23. Donatelli P (2015) Perfectionist returns to the ordinary. MLN 130(5):1023–1039
24. May T (2017) A fragile life: accepting our vulnerability. University of Chicago Press, Chicago
25. Velázquez LG (2020) The role of philosophy in the pandemic era. Bioeth Update 6(2):92–100
26. Furedi F (2004) Therapy culture: cultivating vulnerability in an uncertain age. Routledge, London and New York
27. Debrix F, Barder AD (2009) Nothing to fear but fear: governmentality and the biopolitical production of terror. Int Political Sociol 3(4):398–413
28. Badiou A (2001) Ethics: an essay on the understanding of evil, trans. Hallward P. Verso, London and New York
29. Frawley A (2015) Semiotics of happiness: rhetorical beginnings of a public problem. Bloomsbury, London
30. Jouan M, Laugier S (eds) (2008) Comment penser l'autonomie ? Entre compétences et dépendances. PUF, Paris
31. Furedi F (2008) Fear and security: a vulnerability-led policy response. Social Pol Admin 42(6):645–661.
32. McLaughlin K (2012) Surviving identity: vulnerability and the psychology of recognition. Routledge, London and New York
33. Ecclestone K (2016) Behaviour change policy agendas for "vulnerable" subjectivities: the dangers of therapeutic governance and its new entrepreneurs. J Educ Policy, http://tandfonline.com/; https://doi.org/10.1080/02680939.2016.1219768
34. Butler J (2006) Precarious life: the powers of mourning and violence. Verso, London and New York
35. Cavarero A (2007) Horrorism: naming contemporary violence, trans. McCuaig W. Columbia University Press, New York
36. Tronto J (1987) Beyond gender differences to a theory of care. Signs 12(4):644–662
37. Fineman MA (2010) The vulnerable subject and the responsive state. Emory Law J 60(2):251–275
38. Nussbaum M (2001) The fragility of goodness luck and ethics in Greek tragedy and philosophy. Cambridge University Press, Cambridge
39. Shulman G (2011) On vulnerability as Judith Butler's language of politics: from "excitable speech" to "precarious life." Women's Stud Quart 39(1–2):227–235
40. Adorno TW (2004) Negative dialectics, trans. by Ashton EB. Routledge, London and New York
41. Honneth A (2004) A social pathology of reason: on the intellectual legacy of critical theory. In: Rush F (ed) The Cambridge companion to critical theory. Cambridge University Press, Cambridge UK, pp 336–360
42. Laugier S (2015) La vulnérabilité des formes de vie. Raisons politiques 57(1):65–80
43. Jaeggi R (2018) Critique of forms of life, trans. Cronin C. Harvard University Press, Harvard
44. Miller EP (2002) Bodies and the power of vulnerability: subjectivity outside the logic of confrontation. Philos Today 46(5):102–112
45. Petherbridge D (2016) What's critical about vulnerability? rethinking interdependence, recognition, and power. Hypatia 31(3):589–604
46. Zhang E (2014) Vulnerability, compassion, and ethical responsibility: a Buddhist perspective on the phenomenology of illness and health. In: Tham J, Garcia A, Miranda G (eds) Religious perspectives on human vulnerability in bioethics. advancing global bioethics, vol 2. Springer, Dordrecht, pp 41–52
47. Esposito R (2023) Common immunity: biopolitics in the age of the pandemic, trans. Hanaf Z. Wiley, New York
48. van de Poel I, de Wildt T, van Kooten Pássaro D (2022) COVID-19 and changing values. In: Dennis MJ, Ishmaev G, Umbrello S, van den Hoven J. (eds) Values for a post-pandemic future. Springer Nature, Cham, pp 23–58
49. Clough B (2017) Disability and vulnerability: challenging the capacity/incapacity binary. Soc Policy Soc 16(3):469–481

50. Vallaud-Belkacem N, Laugier S (2020) La Société des vulnérables. Leçons féministes d'une crise. Gallimard, Paris
51. Furedi F (2003) A culture of fear revisited: risk taking and the morality of low expectation. Continuum Books, London
52. Lasch C (1984) The minimal self: psychic survival in troubled times. W.W. Norton, New York
53. Agamben G (2021) Where are we now?: the epidemic as politics. Rowman & Littlefield Publishers, Lanham
54. Polychroniou A (2022) Towards a radical feminist resignification of vulnerability: a critical juxtaposition of Judith Butler's post-structuralist philosophy and Martha Fineman's legal theory. *R*edescriptions: Polit Thought, Concep History Feminist Theo 25(2):113–136

Stefania Achella is Professor of History of Philosophy at the University of Chieti-Pescara (Italy), where she teaches History of Philosophy. She is currently working on the relationship between philosophy and biology in the nineteenth century, on the concepts of life and corporeality in classical German philosophy, and on the contributions of Feminism to the philosophical debate. Among her most recent publications: "Idealism and Science of Life: An Intersection Between Philosophy and Biology" (Springer 2022); "Limit Situations: For an existential anthropology of vulnerability" (Rivista di Studi Jaspersiani 2022); *The Owl's Flight. Hegel's Legacy to Contemporary Philosophy* (co-editor, De Gruyter 2021); "Gendering the Anthropocene?" (Itinerari, 2020).

Chapter 2
Ethics in Scenes of Disaster

Piergiorgio Donatelli

> *The suspicion of the ordinary seems to me to be rooted in the fact that relationships require a repeated attention to the most ordinary of objects and events, but our theoretical impulse is often to think of agency in terms of escaping the ordinary rather than as a descent into it*
> *Veena Das, Life and Words, 2007*

Abstract This chapter delves into scenes of disaster as crucial sites to explore the role of forms of life and the ordinary. The habitual and ordinary have the power to bring life together when a form of life breaks down. However the ordinary is constantly at play in our lives allowing for what we perceive as autonomous agents to make choices. The theory of anthropologist Veena Das is specifically examined in this chapter for the important lessons she reveals from the philosophical tradition of Wittgenstein and Stanley Cavell. The COVID-19 pandemic has shown the importance of shifting attention to the network of interdependencies that gives life to the great values of democratic societies and especially to the invisible work of care lavished by many invisible people. This lesson has not been learned.

Keywords Disasters · Ordinary · Forms of life · Veena Das · Stanley Cavell · Ludwig Wittgenstein

1 Introduction

In what follows I won't be directly discussing the Covid-19 pandemic. I will explore instead how moral and political thought can be reshaped once we pay attention to circumstances where life is preserved and repaired by work hidden in the very common actions performed as a part of the everyday routine. Strands in the ethics of care have shown how the work of care that helps to sustain life in most circumstances

P. Donatelli (✉)
Sapienza Università di Roma, Rome, Italy
e-mail: piergiorgio.donatelli@uniroma1.it

S. Achella and C. Marazia (eds.), *Vulnerabilities*, Integrated Science 18,
https://doi.org/10.1007/978-3-031-39378-5_2

is actually performed by people who we do not truly see and whose importance is not truly acknowledged. However, the power of the everyday and of the ordinary—two signature notions for both Ludwig Wittgenstein and Stanley Cavell—can be especially observed in situations in which the continuation of life, in its biological and social dimensions, is severely endangered, as with mass violence and disasters. Here we may come to acknowledge how what is most ordinary and common is both capable of sustaining life and of hosting what is found to be otherwise inexpressible, the radical ruptures of habits and expectations. Everyday activities that create the routines of life reveal our uncanny habitation in the world. They embody both our natural attachment to people and things as well as our elusion and weariness, as they represent in fact routine and repetition, not the vigorous moments of choice and decision. Wittgenstein's and Cavell's lessons on the ordinary and the work of the anthropologist Veena Das on situations of violence and extreme affliction unearth the power of ordinary life conceived in its dimensions of passivity and vulnerability. Where we expect to find only exposure to adversity and disgrace, we also discover how people uncover the power to incorporate these blows in the ordinary rhythms of life, finding ways to respond to circumstances.

These considerations, which I will expand on in the following chapter, may be drawn in connection to the recent experience of COVID-19. A crucial lesson that can and should be learned from it certainly concerns the importance of science, technology and democracy. The political response to the pandemic has highlighted a synergy between the different branches of science, medicine, digital technologies, artificial intelligence, and the democratic organization of societies. The results were impressive. Scientific research was able to promptly sequence the virus, which was shared through the GISAID initiative on January 10th, 2020. After 10 months, Pfizer-BioNTech submitted a vaccine to the U.S. Food and Drug Administration on November 20th. Approvals from national regulatory agencies quickly followed, and billions of doses were administered worldwide. These results need to be highlighted, especially if we recall the flawed responses that came from governments, such as those of Trump, Bolsonaro, and Modi, who undermined the problem, criticized expert committees, fostered misinformation, and weakened trust in the scientific community. Similar responses were also cultivated by significant political minorities within democratic nations governed by cabinets who enforced rational lines of response to the pandemic.

However, if we focus only on the most visible actors in the successes (leaving aside the failures that are to be found, mostly in the area of international justice as well as in national public health systems), we leave out of account the many people and situations that contributed to keeping our societies functioning, especially in the most severe months of the pandemic. Actors such as scientists, pharma companies, politicians and public health officials could accomplish these outstanding results only in collaboration with a great number of people who worked alongside them in pharmaceutical companies, transportation, and the various infrastructures of society. These and many other people were in the first place in charge of the work that allowed life to continue in viable rhythms for all. In the hard months of the pandemic, there was much talk about the heroes in hospitals, from which as a matter of fact nothing significant followed now that we are in a position to rethink the role of public health in our societies. However, no attention was ever truly given to the many other people, from cashiers to garbage collectors, delivery people, truck drivers, caregivers and those, especially women, who took care of their families, in particular of the young and elderly. This kind of work needs to be put in the foreground both because it allowed life to continue with some normality and because to this normality we could all entrust our inability to take in what was happening, allowing it to be elaborated and transformed into a renewed incentive to face what was happening and broach a novel sense of responsibility. We therefore need a change in what we consider central in ethics and hence in political action, a revision of the categories that would allow us to look at the invisibilities of people and their activities and to give central stage to the network of interdependencies that gives life to the great values of democratic societies, such as personal freedom and justice, scientific research and technological progress.

2 Wittgenstein and Forms of Life

My argument will start with a defense of an approach based on the idea of forms of life as can be found at the heart of Ludwig Wittgenstein's philosophy. Wittgenstein teaches us to read intellectual forms (concepts and rules) in the life where they have a position and connections. At the outset of the *Philosophical Investigations*, he shows how speaking, in the example of calling out an object, is woven into a form of life. In Sect. 2, he offers the example of two builders:

> A is building with building stones: there are blocks, pillars, slabs and beams. B has to pass him the stones and to do so in the order in which A needs them. For this purpose they make use of a language consisting of the words "block", "pillar", "slab", "beam". A calls them out; B brings the stone which he has learnt to bring at such-and-such a call. – Conceive of this as a complete primitive language. [1, §2, p. 6e]

Trying to conceive this situation as a complete language game, albeit a primitive one, involves being able to understand what calling out an object is by looking at what these people do, at its place in this very primitive context of life. As Wittgenstein writes in section 23, "The word 'language-game' is used here to emphasize the fact that the speaking of language is part of an activity, or of a form of life" [1, §23, p. 15e]. In the example of the builders from Sect. 2, the context of life is so scanty that we hesitate to consider the two builders as real human beings, since the entire normal context of life all around what they are doing and pervading their actions is missing (they seem more like robots or mechanized human beings) [2, pp. 124–186]. This hesitation indicates a more general difficulty in looking at the life of the use of a sign. Wittgenstein writes that we should be wary of the inclination to look somewhere else from where a linguistic sign is used, which he glosses as an inclination to sublimize language [1, §38, pp. 22e–23e]. We are asked instead to read naming and any other linguistic activity and intellectual form in the life in which they are woven as recurrent patterns in the weave of life. We are accustomed to considering our concepts, rules and principles as rigid, as fixed rails along which our thinking runs. This attitude accounts for the normative role they play in our lives, guiding us, say in responding to the word slab with the action of bringing the object slab to the person who is calling it out, in Wittgenstein's example of the builders, or in continuing a numerical series (2, 4, 6, 8…). This normativity, which envelops us with a sense of constriction, does not derive from some authority transcending the actual life where it plays its role. Such authority is, on the contrary, the result of a fitting together of aspects, the holding together of the many details that go into building a form of life, the life with calling out an object such as pillars and slabs or the life with numerical series.

As Wittgenstein writes,

> Why do we call something a "number"? Well, perhaps because it has a – direct – affinity with several things that have hitherto been called "number"; and this can be said to give it an indirect affinity with other things that we also call "numbers". And we extend our concept of number, as in spinning a thread we twist fibre on fibre. And the strength of the thread resides not in the fact that some one fibre runs through its whole length, but in the overlapping of many fibres. [1, §67, p. 36e]

There is no external, transcendent, authority establishing what is to be called a number: the normative force of the concept of number is similar to the spinning of a fibre, where the force of the fibre is not derived from some rigid mechanism running through it but from the overlapping of the many fibres. Similarly, the normativity of intellectual activities can be seen as a form, a pattern, emerging in the weave of life. It is "we, in our conceptual world, [that] keep on seeing the same, recurring with variations" [3, §672, p. 115] sorting out aspects based on what is important to us [3, §638, p. 110].

It is crucial to Wittgenstein's approach that reading a form into a context of life comes with seeing this form in life, overcoming the inclination to look for two separate items, form and life: that is, the intellectual activity equipped with its normative force, on the one hand, and a certain description of how people behave and the facts of the matter that concern them, on the other. The temptation Wittgenstein is trying to overcome is to consider the two apart and then register the correlation between them. As Wittgenstein writes in *Zettel* 351:

> If humans were not in general agreed about the colours of things, if undetermined cases were not exceptional, then our concept of colour could not exist." No: – our concept would not exist. [4, §351, p. 64e]

Cora Diamond has worked out the point very nicely. She writes that Wittgenstein is rejecting.

> the idea that there being a complex life with colour terms, a life involving agreement, is one thing, and that our having our colour concept is *something else*, standing or not standing in a relation of logical or conceptual dependence to that complex life involving agreement. If you think of the complex life as something else, you might say that people *could* not have our colour concept unless they had that complex life into which agreement enters. What Wittgenstein replies is that if they did not agree about the colours of things, and hence did not have that complex life but lived in some quite different way, that would *be* their not having our colour concept. [5, p. 19]

The difficult task is to describe a form of life as what gives life to our various intellectual activities without retaining the sense that their normativity (the normativity of arithmetic and of ordinary concepts such as colors, tables and flowers, for example) survives in some strange way what happens to the life where they are ingrained. At the same time, seeing intellectual forms in the life where they have a role does not disperse the logical force, as it were, in their life: the logical force of adding two in a numerical series or the appropriateness of the description of the colors irradiating in a sunset is still there, constraining what we say, yet it is there in the form of life where numerical series and the application of color concepts have a place. By reading intellectual activities in the life where they have a place and connections, we are not destroying their logical force; we are not reducing it to something else, to external conditions that explain its constraint in different terms. Wittgenstein's appeal to forms of life is not a way of doing away with the normative force of intellectual forms or of concepts and rules. It is not a reductionist program translating logical force into social and natural considerations, although he was sometimes interpreted as such an author. Nevertheless, things are more complicated than this because in exploring life with our concepts and rules, these are transformed after the exploration.

3 Exploring Forms of Life in the Ethics of Care

If we turn now to ethics, we can similarly try to treat ethical thought as the family of forms that can be read in the moral life of people. Important strands in moral philosophy have made this very difficult, however. The novelist and philosopher Iris Murdoch was someone who argued convincingly that an influential picture of the moral subject depicts moral agency as concentrated in the specific moments of choice and in the deliberation required to support choice with appropriate reasons, thus leaving out completely the life that goes on continuously in the background of the specific episodes singled out as the central scene of morality. The picture she addresses is the following.

> On this view, the moral life of the individual is a series of overt choices which take place in a series of specifiable situations. [...] Further, a moral judgment, as opposed to a whim or taste preference, is one which is supported by reasons held by the agent to be valid for all others placed as he, and which would involve the objective specification of the situation in terms of facts available to disinterested scrutiny. [...] The charms of this view are obvious. It displays the moral agent as rational and responsible and also as free; he moves unhindered against a background of facts and can alter the descriptive meaning of his moral words at will. [6, pp. 77–78]

According to this picture, we all live in the same empirical and rationally comprehensible world. However, this hides from our view the fact that in a different sense, we live in a world shaped by the many aspects that are visible in the ramifications of concepts "which themselves determine a vision of the world" so that "the prohibition on defining value in terms of fact loses much of its point" [6, p. 94]. On this different perspective of morality and agency, Murdoch argues,

> freedom here will consist, not in being able to lift the concept off the otherwise unaltered facts and lay it down elsewhere, but in being able to "deepen" or "reorganise" the concept or change it for another one. On such a view, it may be noted, moral freedom looks more like a mode of reflection which we may have to achieve, and less like a capacity to vary our choices which we have by definition. [6, p. 95]

Murdoch is thus encouraging the exploration of a life that goes together with moral deliberation and choice and argues that this would be a work that shows the different forms moral thought and agency take in the contexts of life.

A good example of what this exploration would amount to is offered in the area of the ethics of care. I will say something briefly about this and then move on to my central concern in this chapter, which concerns situations of disasters and of mass violence where a form of life has fallen apart.

Strands in the ethics of care have shown how the idea of the autonomous subject needs to be rethought. The paradigm centered on justice requires us to occupy the impartial point of view to recognize each person's needs and demands and to respond to them appropriately. Seen from the perspective of the ethics of care, this way of proceeding entirely misses the work that needs to be done to recognize the demands of others and our responsibilities as well. One must not turn away from where one is situated within networks of relationships but instead must deepen that sense and bring to light one's responsibility for the conditions that make people's lives burdensome or dramatic. It is not so much a matter of seeing oneself from the rational point of view from above but of recognizing how one's fortunate condition depends, for example, on the labor lavished by people without it being considered or having it hidden in the form of economic relationships; it is a matter of coming to terms with one's condition of privilege, recognizing one's dependence on others. Deepening this network of relationships reveals our involvement in conditions of exploitation, hardship, poverty and unhappiness. These situations emerge for us as sources of concern and of moral demands within the relationships that connect us to these lives: relationships that we have grown by "deepening" our place in the world, to go back to Murdoch's suggestion in the quotation above. Care is what sustains life, Joan Tronto writes [7, p. 117], and it is the care poured in order to preserve human forms of life that offers a perspective that reveals responsibilities and needs situated in specific relations even though they are systemic and global. The perspective of justice, by contrast, can easily hide individual responsibilities behind descriptions of reality where the privilege of some and the exploitation of others appear in the morally neutral terms of the functioning of the economic system.

My comment is the following. Pursuing the work suggested by Murdoch and the ethics of care we do not do away with the idea of individual thought and deliberation, of agency and responsibility, but nevertheless the face of these concepts changes, the individual does not disappear yet the sense of what she does is transformed. By looking at what is not the object of choice but is silently assumed as the normal conditions that enable agency, the notions of freedom and responsibility are transformed. The ethics of care is especially congenial in showing various sorts of invisibility. There are many invisibilities that help some to flourish and that leave other people deprived of such possibilities and voiceless. These invisibilities also affect the privileged, as they deprive them of the knowledge of others and of themselves. They can be seen, and they may come to see themselves, as prisoners of their privilege and of the blindness inherent in it. James Baldwin has made a clear case of this problem. He writes:

> I have always been struck, in America, by an emotional poverty so bottomless, and a terror of human life, of human touch, so deep that virtually no American appears able to achieve any viable, organic connection between his public stance and his private life. This failure of the private life has always had the most devastating effect on American public conduct, and on black-white relations. If Americans were not so terrified of their private selves, they would never have become so dependent on what they call the 'Negro problem.' This problem, which they invented in order to safeguard their purity, has made of them criminals and monsters, and it is destroying them. [8, pp. 53–54]

To return to our initial problem, in these treatments, the scene of moral agency is not dissolved in the larger networks of relations; it is still there, but its face has changed. On the one hand, agency has been transformed by showing the structures of interdependency: one's bald sense of freedom and independence has been located within relations from the perspective of which the picture of pure freedom seems inadequate and the weight of one's responsibility has correspondingly grown. Furthermore, agency has migrated to a number of situations and people who were unseen, to the people who are doing a work that sustains life and who are described in a way that hides their agency as the ordinary maintenance of the living, absorbed in the habitual rhythms of life. An exploration of this kind changes our perception of the people involved, as we see them as concerned in relations where their agency is hidden in what is otherwise perceived as the normality of life and their lives are in this manner impoverished and impaired. This background left in the dark becomes a scene of moral importance.

4 Ordinary Life in Spaces of Devastation

The kinds of transformations and reallocations of matters of importance I have just discussed are particularly visible in the conditions of the catastrophic collapse of forms of life. These are compelling situations for exploring and rethinking established ways of approaching ethics. In this context, the analysis offered by Veena Das is of particular relevance. She is an anthropologist and a philosopher, taking her lead from Wittgenstein's and Stanley Cavell's work. She shows what is wrong in a number of approaches that freeze what happens when a whole form of life collapses in a scene of pure victimization. She discovers the occasions for moral response and a live sense of agency where one would be inclined to see none.

Veena Das is keenly aware of the difficulty of describing the normal course of life without idealizing it or deforming it. An inclination in philosophy, also to be found in anthropology and in public conversation more generally, is that of treating the ordinary as a place of unredeemable normality or misery, as an otherness that haunts us and that may be treasured only in the form of negative critique. The risk that is being run by this is to aestheticize the ordinary by turning it into a source of interesting pleasures that enrich our lives, sealed from what could put them in jeopardy: the risk inherent in the familiar phenomenon of spectacularizing catastrophic events. One example comes from the discussion in her book *Slum Acts* of the Bombay blasts in 1993. I will refer to a comment she makes about an article on Bombay's Black Friday written by the anthropologist Vyjayanthi Rao. Rao writes: "The city appeared without individuality or particularity—without boundaries, without the recognition of an inside or an outside, of an enemy, or another, but as a pure instance of victimhood. It was, as well, a specifically global moment, for the city as it joined what were the areas similarly targeted by acts of terrorist violence" [9, p. 571; mentioned in 10, p. 40]. Das's comment is that "Rao opts for capturing what middle-class residents might have felt as currents of feelings without identifiable agents or subjects swept the city" [10, p. 41]. Das captures the kind of setting that enacts the peculiar sentiment of the sublime in Kant's Third *Critique*, exemplifying our happy return to our own transcendental subjectivity, which leaves no scars on our body after the contemplation of the perilous storm stirring the sea. This has epistemological and political consequences, however, which Das draws clearly. She writes, "The 'global moment' in which [Rao] places the city as 'pure victim' elides the real presence of the chawls and the low-income localities from where most of the accused came" [10, p. 42]. The slum acts, their language and the very life of individual people and communities are deleted from the scene. Das observes that uncanny resonances can also be heard between police language and this scholarly prose, as it was precisely this construction of global terror, revenge, unknown perpetrators and the city portrayed as a mere victim that was functional to the actions taken by the police, extending the period of custody for the accused and creating the story of Muslim hurt and revenge instrumental to Muslim persecution [10, pp. 41–42].

Das explores the ordinary reality of this condition of victimhood and unveils how life is maintained in everyday activities such as preparing food, taking care of the home, and looking after people. Pure victimhood is transformed into a rich scene of acts and language that is torn apart by violence but that leaves its mark in gestures and silences. The perspective of the ordinary shows how life continues in the midst of extreme violence in a situation that is both experientially numbed and that also hosts all these normal activities. This is a central theme in her book *Life and Words*. Working on the acts of violence that followed the Partition in 1947 and those in the aftermath of the assassination of Indira Gandhi, she writes that "my engagement with the survivors of riots also showed me that life was recovered not through some grand gestures in the realm of the transcendent but through a descent into the ordinary" [11, p. 7]. This ordinary is especially enacted by women who in the midst of riots took care of the details of everyday life, such as preparing food, looking after children, washing and keeping the house in order. What appears here is important for my argument about the moral subject. As Das writes on the very same page of *Life and Words*:

> The suspicion of the ordinary seems to me to be rooted in the fact that relationships require a repeated attention to the most ordinary of objects and events, but our theoretical impulse

> is often to think of agency in terms of escaping the ordinary rather than as a descent into it. [11, pp. 6–7]

This recoil from the ordinary produces the picture of the moral subject that gains independence from the life where it has a position and connections.

In *Slum Acts,* Das explores the practice of police investigations after the Bombay blasts and especially the story of a man, Abdul Wahid Shaikh, who finds his personal way of enduring torture and humiliation at the hands of the police by investing these acts with meaning and especially through his writing [10, pp. 66–67]. Shaikh was detained in prison for almost a decade before being acquitted of all charges, and he wrote the book while he was in prison, threatened and tortured by the police officers. There is a human texture that knits together the acts of torture perpetrated by the police officers, their procedures, the details of the places where he was detained, and Shaikh's perseverance, glossed by Das as a political response, but which is also an act of personal resistance and transformation. As Das writes,

> If Wahid Shaikh is able to show some mastery over the experience of torture, it is not that he is somehow healed, but that the impulse to not let his experience disappear is simultaneously steeped in a cultural imaginary (say that of Allah watching what is happening to you) and an attempt to convert a form of shaming that comes from the experience of the human (an animal would not be humiliated by a lack of clothes) to another register in which the mark of humiliation is to be converted into a sign of not giving up. [10, p. 66]

The space of active responses offered by Shaikh is made invisible in the language of legal procedures used in courts, and it is also absent from the descriptions offered by the social sciences that stop short at what appears as simply unspeakable cruelty. Another example treated in the book is the story of an 8-year-old girl (referred to as Kh) who was abducted, raped and tortured in the Delhi area. Das comments as follows:

> as we saw in the cases of torture documented by Wahid Shaikh, there was an uncanny proximity of the physical torture and the constant flow of ordinary talk by ordinary police inspectors. Even in the case of torture inflicted on Kh, it was embedded in the routine acts of cooking and entrusting Kh with the task of looking after the two children of the abductor who resided in the same room. [10, p. 112]

In these stories, we can see how someone like Shaikh finds his way to respond to a situation in which he is otherwise completely victimized. A particular figure of the moral subject emerges in this exploration. It is not the kind of subject criticized by Murdoch because practical resources are found precisely in relationships with reality that are not chosen: they are not the result of free choices. Choices are actually made by the protagonists of Das's ethnographic work against the background of involvements dominated by affliction, passivity and victimization. However, it is precisely when viewed through such descriptions that these people become capable of mobilizing themselves to respond to what is happening to them, simply by being able to tune the violence into the rhythms of everyday activities or, as with Shaik's story, by taking in the violence suffered and converting it through his writing into moments of self-learning.

5 Experience at Its Limits

Our understanding of ethics has changed after these anthropological explorations. We are ready to register occasions of choice and freedom where previously we wouldn't have noticed any. The picture of the moral subject that was the target of Murdoch's criticism does not work because it abstracts ethics from its life. By exploring this life, many aspects change. The idea of deliberation changes. We need to be ready to see aspects of importance where we would not have expected to find them. The knowledge required is inordinate, writes Das [10, p. 115]. We do not know in advance what may count as an occasion for critical response: reflective resources are not laid out in advance. The shape of freedom and the opportunities for choice, the resources to employ to form goals, are all to explore. We must look at details, as Wittgenstein writes: "In order to see more clearly, here as in countless similar cases, we must look at what really happens in detail, as it were from close up" [1, §51, p. 30e]. In Sect. 3, when I introduced the ethics of care, we saw how our freedom is based on a network of interdependencies that transform the picture of free choice. Our freedom is found against the background of the rhythms of life, of a repetitive everyday, where work is done in conditions that are not those of choice. This repetitive everyday is turned into a scene of minor acts and gestures of care that sustain life for all. In Das's examples, this repetitive everyday can host the violent ruptures that she recounts in her work. How can it do this and how does it host it?

The conditions are those in which violence has the power to numb language, life is conducted as in a trance. Language is deadened by violence, yet this experience is lived and channeled in the rhythms of the everyday care of life. Anthropological work (work on the forms of life) unveils these moments, when choice, thought and agency are experienced at their limits. The ordinary appears after Wittgenstein and Cavell as this sort of place.

Experience at its limits is a Wittgensteinian theme and a modernist one as well, which we can spell out as follows. The impossibility of undergoing violence and destruction and responding to it critically and actively is experienced in the form of a trance. This lived impossibility takes the shape of experience being severed from its context. Wittgenstein's *Philosophical Investigations* portrays language losing its foothold in our lives, displacing us on the surface. We experience life on the surface of words, which are enjoyed in their secondary meanings. This is also the issue of Central European modernists, such as Kafka and Gombrowicz: in the works of these writers, worlds are constructed with the internal forces of literary writing; meaning is created immanently within the world constructed with language. A world is created on the surface of language.

I have only gestured at a very complex theme, pointing to the occasions when expressive resources are wearied, and language and life seem capable of clinging to the other only at their surfaces since the bulk of the connections to the world is lost. Modernists thought that this was a condition of both exhaustion and imperceptible creativity. Let us now go back to Das and to the scenes of disaster. Following Cavell's lead, Das writes that the everyday:

> is both a space of routines and habits as well as a space that contains the potential of generating the kind of doubts about one's relations to others and to oneself that can become world annihilating. This feature of everyday life is what gives it an uncanny character. [12, p. 58]

The catastrophic event has the power to annihilate the texture of human connectedness to the point where people are haunted and become ghosts of themselves, like automata. The German writer and literary critic W.G. Sebald is a very important author in this connection. The concept of haunting life by inhabiting it like a ghost is important for him. It is one of the themes in *Austerlitz*, for example, when Austerlitz contemplates the photo of himself as a child dressed like a page, overwhelmed by the sense of the long years that had passed. Austerlitz explores the habitation of the present as itself infected by this uncanny sense of the past gazing at us from the magic halo of a photo:

> It does not seem to me, Austerlitz added, that we understand the laws governing the return of the past, but I feel more and more as if time did not exist at all, only various spaces interlocking according to the rules of a higher form of stereometry, between which the living and the dead can move back and forth as they like, and the longer I think about it the more it seems to me that we who are still alive are unreal in the eyes of the dead, that only occasionally, in certain lights and atmospheric conditions, do we appear in their field of vision. As far back as I can remember, said Austerlitz, I have always felt as if I had no place in reality, as if I were not there at all, and I never had this impression more strongly than on that evening in the Šporkova when the eyes of the Rose Queen's page looked through me. [13, p. 185]

Sebald works on scenes of destruction with a sensibility for the turns of language in such circumstances. He explores the theme of the bombings of German cities in World War Two. In his comments on the work of the German author and film director Alexander Kluge, he points to examples of how in desperate situations people go on with the same everyday occupations and are totally unprepared to face the catastrophe that is being announced, such as the lady employed at the cinema in Halberstadt after the bombing in 1945, who cleans up the rubble for the afternoon movie showing and clears the cellar by dumping the charred body parts in the washhouse boiler; or the women on watch in the tower equipped with folding chairs and thermos and sandwiches still reporting as the tower is falling down, and other stories like these. Sebald offers the following comment:

> These and many of the other stories making up the text show how, even in the middle of the catastrophe, individuals and groups were still unable to assess the real degree of danger and deviate from their usual socially dictated roles. [They show] that the autonomy of mankind in the face of the real or potential destruction that it has caused is no greater in the history of the species than the autonomy of the animal in the scientist's cage, a circumstance that enables us to see why the speaking and thinking machines described by Stanisław Lem wonder if human beings can actually think or are merely simulating that activity, and drawing their own self-image from it. [14, pp. 89–90]

In these examples, habitual behavior becomes automatic, and human beings turn into ghosts or copies of the real individuals. These situations are such that action, speech and gestures are deprived of the substance of life, as really normal life has been suppressed and habits survive haunting the world like phantoms, when normal human behavior turns horrible and creepy.[1] Here is another quotation from *On the Natural History of Destruction*:

> Nossack describes seeing a woman cleaning the windows of a building "that stood alone and undamaged in the middle of the desert of ruins... We thought we were looking at a madwoman," he writes, and continues, "We felt the same when we saw children tidying and raking a front garden. It was so far beyond all comprehension that we told other people about it, as if it were some sort of a marvel. One day we came to a suburb that had not suffered at all. People were sitting out on their balconies drinking coffee. It was like watching a film; it was downright impossible." Nossack's sense of alienation arose from seeing himself confronted, as it must have seemed from the viewpoint of one affected, by a lack of moral sensitivity bordering on inhumanity. You do not expect an insect colony to be transfixed with grief at the destruction of a neighboring anthill, but you do assume a certain degree of empathy in human nature, and to that extent there is indeed something alarmingly absurd and shocking about continuing to drink coffee in the normal way on Hamburg balconies at the end of July 1943, rather like the sight of Grandville's animals, in human dress and armed with cutlery, consuming a fellow creature. [15, pp. 41–42]

The pertinence of the notion of the everyday is to be found in these neighborhoods. The everyday is the habitual and the repetitive that can turn into the uncanny and trancelike, representing this sort of estrangement from humanity, this condition of skepticism. The habits and patterns of life that have estranged themselves from their ordinary context and that appear uncanny still retain a sanity of their own though, as Sebald remarks in the next lines from the quotation I have just offered:

> On the other hand, keeping up everyday routines regardless of disaster, from the baking of a cake to put on the coffee table to the observance of more elevated cultural rituals, is a tried and trusted method of preserving what is thought of as healthy human reason. [15, p. 42]

Habits that have become trancelike and uncanny still retain the power to knit life together again. Stanley Cavell offers a remark about Charlie Chaplin's *The Gold Rush,* which is pertinent in this context. He comments on

> two most famous set routines from that film, the Thanksgiving dinner of roast shoe, and the dream-dance of the rolls on forks. [...] in the one case a shoe is treated as a food (a case of dire necessity), in the other a food is treated as a shoe (a case of dire luxury); in both his imagination gives habitation to his ecstasy and to his grief. The madness of his meaning keeps him sane. [16, p. 176]

The last observation is important in the context of our argument: the madness of Chaplin's imagination is what keeps him sane.

---

[1] I should also recall Wittgenstein's treatment of this example in 1, §420, p. 133e, "But can't I imagine that people around me are automata, lack consciousness, even though they behave in the same way as usual? If I imagine it now a alone in my room I see people with fixed looks (as in a trance) going about their business the idea is perhaps a little uncanny. But just try to hang on to this idea in the midst of your ordinary intercourse with others in the street, say! Say to yourself, for example: 'The children over there are mere automata; all their liveliness is mere automatism.' And you will either find these words becoming quite empty; or you will produce in yourself some kind of uncanny feeling, or something of the sort. Seeing a living human being as an automaton is analogous to seeing one figure as a limiting case or variant of another; the cross-pieces of a window as a swastika, for example".

The impossibility of experiencing created by catastrophic events produces the situation in which the forms of experience become uncannily independent and isolated, the issue modernists such as Kafka and Gombrowicz tied to an analysis both of society and of the human condition and that we find intimated in the treatment of the sites of mass destruction and violence offered by Das and Sebald. The language and gestures inhabiting the surface of experience are the only expressive possibilities left for these people; they represent one face of the everyday having become a ghost of itself, the other side being the kind of work this drained everyday can still achieve. As with Chaplin, "The madness of his meaning keeps him sane".

Das writes in *Life and Words*:

> Indeed, if we persist in defining the subject primarily through her relation to knowledge, then we would have to move in the direction of showing how the subject is *only* constituted through loss. In the literary rendering of the Partition, as we saw in chapter 3, this loss was seen as the inability of the women to find a way of telling their story as part of the story of the nation – but in their small communities defined by everyday relations, women were able to redefine themselves through the work of repair that they performed. So one has to understand not only the themes of loss and concealment and the almost hallucinatory quality of the speech generated in the riots but also the themes of how one might shift one's gaze to the inhabitation that comes not from the knowing subject but from the subject as engaged in the work of stitching, quilting, and putting together relationships in everyday life. [11, p. 161]

Das contrasts here two perspectives, one represented by the knowing subject and one occupied by the subject engaged in the ordinary activities that knit life together. My reading of Das is that we can draw together the two descriptions and consider "the almost hallucinatory quality of the speech" as one side of the inhabitation of the subject "engaged in the work of stitching, quilting, and putting together relationships in everyday life". The hallucinatory quality of speech is a way in which a life of affliction and violence can be endured and knitted together once again. Such hallucinatory uses of language can be the only way to preserve life. Similarly, we could argue that Kafka thought that the hallucinatory constructions in his stories were the only ways a haunted life could be brought together and lived, and Gombrowicz thought that the construction of a private mythology out of the proper mature Form is the only way to reinhabit one's ghostly life.

Radical violence and loss are experienced in ordinary reality, and this experience is made possible through ordinary actions and words, the habitual actions described by Sebald, the ordinary activities and talks intermingled with torture and violence in the stories of Shaik and Kh and of the people in the riots after the Partition in India described by Das. These moments of ordinary life can thus play the double role of continuing life after destruction and of absorbing this ghostly inhabitation of the world. They can do both because the ordinary is such a place of routines and habits and of skeptical flight from reality.

This is a peculiarly Wittgensteinian strand that concerns the status of language being always on the verge of losing itself while we lose ourselves in language, away from home and on its way of being led home, as in the famous passage from the *Philosophical Investigations* 116: "What *we* do is to bring words back from their metaphysical to their everyday use" [1, §116, p. 53e]. This inherent ambivalence of language is also the ambivalence of our ordinary life, of what Cavell calls the domestic, the normal rhythms of life from day to night, from season to season, from year to year, a repetition which installs a continuity and a certainty, a trust in oneself, in others, and in the world. This certainty and trust are infected by skepticism, by the susceptibility of our ordinary dealings to inhabit the world as ghosts, as automata, which is the Kafkian theme also commented on by Sebald. However, this skepticism can turn into a dimension of transformation and recovery. This ghostly habitation is on the one hand necessitated as we are driven from our world by violence and torture, by the breaking apart of a form of life. However, this state of passivity, e.g., when we are in the hands of torturers as in Shaik's story, can be turned into an occasion to respond and engage with reality. The ambivalence of words and actions can explain these kinds of resources, which are not seen if we abstract from such material—this inordinate knowledge as Das calls it, following Cavell—as we do when we work with the picture of the moral subject criticized by Murdoch.

The appeal to the space of morality and responsibility found in the circumstances in which we have not chosen anything and we cannot be properly pictured as free and autonomous agents in the Kantian way, or in the social contract framing of morality, are usually employed to undermine the importance of freedom and individuality. (In bioethics, e.g., the argument is usually made about the areas of life considered as given and structured by an order that lies beyond our human reach). In the specific Wittgensteinian (and modernist) context, the thought takes on a different aspect, and the exploration of situations of passivity in circumstances of mass violence and affliction actually uncovers occasions of creative and critical response, "recovered not through some grand gestures in the realm of the transcendent but through a descent into the ordinary"—to go back to Das's words [11, p. 7]. The larger philosophical picture is that of an intellectual and practical creativity ingrained in the detailed texture of life, in the forms of life, defined through their vulnerability to loss and exhaustion, which can show though a power to repair and mend life again.

6 Conclusions

I started with Murdoch's picture of the moral subject who freely choses against the background of a world of facts she knows and on which she can intervene. Now, at the end, we have moved to a different picture.

We have transformed our notion of agency as we are prepared to find it in the many ordinary activities that sustain life. These ordinary activities take on the aspect of being merely habitual, and the first move is to see them as places of agency and morality: thus, we mobilize what is normally taken as the grey background of life, and in doing this, we also revise the notion of what is normal and taken as a matter of course. Agency is to be found in the very conditions in which people's actions are apparently passive and unreflective. This redescription is radicalized in the analyses of disasters, mass violence and destruction. In Das's examples, creative spaces for action are found within states of victimization and passivity. Freedom and creativity are discovered in the entanglements of the forms of life, not in the crystalline purity of a scene where there are no obstacles and no hindrances.[2] We inhabit our life, conditioned and grounded, yet in having found ourselves somewhere, in the acknowledgement of this condition of passivity, vulnerability and exposure, we may also locate the power to use the materials that make up our passivity as occasions for change and transformation.

The experience of the COVID-19 pandemic with its successes and failures has offered a lesson about the importance of shedding light on the invisible network of interdependencies on which social and individual life were able to continue with some normality. Governments relied on this normality to pursue health policies, and to this normality, we all entrusted our difficulty dealing with the emergency and the unexpected. The COVID-19 pandemic has shown the importance of shifting attention to the invisible work of care lavished by many people and the importance of knitting together life through ordinary gestures of care and trust. However, this lesson has not been learned. We are not registering a renewed prominence of public health systems or of the work of care. The striking consequences that we are registering do not concern the improvement of social infrastructures but the increasing role of private Big Tech companies and excellence in health care against the importance of strengthening the social texture of public health systems and care work. This missed lesson belongs, however, to the problem of the ordinary and to our inclination to deny the vulnerabilities and the network of interdependencies that define us.

Core Messages

- We need to revise the notion of agency to disclose the many invisible people in charge of the work that allows life to continue in viable rhythms for all.
- We mobilize what is normally taken as the grey background of life, and in doing this we also revise the notion of what is normal and taken as a matter of course.
- The COVID-19 pandemic has shown the importance of shifting attention to the invisible work of care lavished by many people and the importance of knitting together life through ordinary gestures of care and trust.

References

1. Wittgenstein L (2009) Philosophical investigations, trans. Anscombe GEM, Hacker PMS, Schulte J. Wiley-Blackwell, Oxford

[2] See 1, §107, p. 51e: "The more closely we examine actual language, the greater becomes the conflict between it and our requirement. (For the crystalline purity of logic was, of course, not something I had discovered: it was a requirement.) The conflict becomes intolerable; the requirement is now in danger of becoming vacuous. We have got on to slippery ice where there is no friction, and so, in a certain sense, the conditions are ideal; but also, just because of that, we are unable to walk. We want to walk: so we need friction. Back to the rough ground!".

2. Cavell S (1995) Notes and afterthoughts on the opening of Wittgenstein's Investigations. In: Cavell S (ed) Philosophical passages: Wittgenstein, Emerson, Austin, Derrida. Blackwell, Oxford, pp 124–186
3. Wittgenstein L (1998) Remarks on the philosophy of psychology, trans. von Wright GH, Nyman H. Blackwell, Oxford
4. Wittgenstein L (2007) Zettel, trans. Anscombe GEM, von Wright GH. University of California Press, Berkley-Los Angeles
5. Diamond C (1989) Rules: looking in the right place. In: Phillips DZ, Winch P (eds) Wittgenstein: attention to particulars: essays in honour of Rush Rhees (1909–1989). Palgrave MacMillan, London, pp 12–34
6. Murdoch I (1998) Vision and choice in morality. In: Murdoch I (ed) Existentialists and mystics: writings on philosophy and literature. The Penguin Press, New York, pp 76–98
7. Tronto JC (1993) Moral boundaries: a political argument for an ethic of care. Routledge, London
8. Baldwin J (2007) No name in the street. Random House, New York
9. Rao V (2007) How to read a bomb: scenes from Bombay's black Friday. Publ Cult 19(3):567–592
10. Das V (2022) Slum acts. Polity Press, Cambridge
11. Das V (2007) Life and words: violence and the descent into the ordinary. University of California Press, Berkley-Los Angeles
12. Das V (2020) Textures of the ordinary: doing anthropology after Wittgenstein. Fordham University Press, New York
13. Sebald WG (2001) Austerlitz, trans. Bell A. Random House, New York
14. Sebald WG (2005) Campo Santo, trans. Bell A. Random House, New York
15. Sebald WG (2003) On the natural history of destruction, trans. Bell A. Hamish Hamilton, London
16. Cavell S (1984) What becomes of things on film? In: Cavell S (ed) Themes out of school: effects and causes. North Point Press, San Francisco, pp 249–257

Piergiorgio Donatelli is Professor of Philosophy at Sapienza Università di Roma. He has written on the history of ethics, contemporary moral theory, bioethics and on issues related to human life, as well as on Wittgenstein, Stanley Cavell, J. S. Mill, and Foucault. Among his recent publications: *Etica. I classici, le teorie e le linee evolutive* (Einaudi, 2015); *Manières d'être humain. Une autre philosophie morale* (Vrin, 2015); *La filosofia e la vita etica* (Einaudi, 2020); *The Politics of Human Life. Rethinking Subjectivity* (Routledge, 2021).

Chapter 3
Humanity of the Human and the Politics of Vulnerability

Rita Fulco

> *There is in each human being something sacred. But it is not his person, which is not anything more than his personality. It is him, this man, wholly and simply.*
> *There is a passerby in the street who has long arms, blue eyes, a mind where thoughts are swirling that I know nothing about, but that may well be nothing special.*
> *It is neither his person nor his personality that is sacred to me. It is him. Him as a whole. Arms, eyes, thoughts, everything. I would not violate any of this without infinite scruples*
> *Simone Weil, Human Personality, 1943*

Abstract What I propose is to explicit the relationship between two dimensions, which, in my opinion, are inseparable from any discourse about vulnerability. On the one hand, the horizon of politics and institutions, the horizon of practice; on the other, that relative to the definition of the 'humanity of the human', the theoretical horizon. To talk about the 'politics of vulnerability', we need to understand what is meant by vulnerability and the dimensions it implies. In this particular case, I want to discuss two philosophers who found a way to keep these two dimensions together, namely, Judith Butler and Simone Weil. The thesis I would like to arrive at could be articulated in the following points: (1) vulnerability is 'constitutive' of the humanity of the human; (2) vulnerability also stems from certain discourses of power; (3) to accept vulnerability as the common trait of humanity can be the basis upon which to construct a nonviolent or least violent possible coexistence; (4) in order to arrive at this, the philosophical and political problems to be addressed are 'attention for' and 'recognition' of vulnerability, and this implies circumstances where vulnerability is not obvious or is not recognised as such.

Keywords Politics · Vulnerability · Humanity · Bare life · Simone Weil · Judith Butler

---

R. Fulco (✉)
Dipartimento di Civiltà Antiche e Moderne (DiCAM), Università degli Studi di Messina, Messina, Italy
e-mail: rita.fulco@unime.it

S. Achella and C. Marazia (eds.), *Vulnerabilities*, Integrated Science 18,
https://doi.org/10.1007/978-3-031-39378-5_3

1 Introduction

What I propose is to explicit the relationship between two dimensions, which, in my opinion, are inseparable from any discourse about vulnerability. These two dimensions demand closer attention, in particular in the wake of the pandemic. On the one hand, the horizon of politics and institutions, that is, the horizon of practice; on the other, that relative to the definition of the "humanity of the human", the theoretical horizon. In order to talk about the "politics of vulnerability", we need to understand what is meant by vulnerability and the dimensions it implies. *In primis*, I would say those of identity, subjective or social.

In this particular case, I want to discuss two philosophers who found a way to keep these two dimensions together, namely, Simone Weil and Judith Butler. For both, the concept of vulnerability is related to the understanding of the humanity of human being.

2 *Affliction and* "Life Without Form"

I am convinced that we need to profoundly rethink the relationship between the humanity of the human and vulnerability to understand whether and how a certain conception of vulnerability can help us transform the juridico-political dimension that we find ourselves confronted with today. For Simone Weil, the human being is *intrinsically* vulnerable, and this vulnerability is the expression of what Levinas defined as "the humanity of the human". Both ethics and politics should be reconsidered starting from this premise. Although a constitutive trait of the human, for Judith Butler perhaps above all, vulnerability is a condition determined, in a Foucauldian manner, by a certain order of discourses. These initial points already allow one to understand that theoretically and politically, both Weil and Butler's philosophical proposition is situated in contrast to Hobbes' *homo homini lupus est* or Thucydides "one commands where one has power"—theses in turn fundamental, for example, for the philosophico-political reflection of a philosopher and jurist such as Carl Schmitt.

To fully understand the thoughts of Simone Weil on these issues, it is useful to recall the spatiotemporal horizon in which she lived and wrote. To recall that she died in 1943 is to underline that all of her works are situated in the period prior to the end of the Second World War and thus that her works are separated from postwar reflections on power, sovereignty, governmentality and biopolitics. This fact helps us better understand (and appreciate) her originality.

To address the issue of vulnerability in Weil, we could pursue several different paths: that of following her first works dealing with oppression in factories, through her reflections on Nazi Germany, to works spanning the years following her exile from France after the Nazi occupation. My aim here is to follow the trace of one concept that unites these different periods, as it names the human condition to which Weil paid the greatest attention: that of 'affliction'. The latter indicates a profound state of psychological and social despondency due to adverse conditions in which any human being can find itself. Simone Weil understood that important philosophical and political questions gravitate around the *affliction* [*malheur*] and grasped its theoretical potential: "*Malheur*, admirable word, without its equivalent in other languages. We haven't got all we could out of it"[1] [1, p. 3]. Affliction [*malheur*] in fact could be identified precisely as 'pure exposure to power': such an exposure is the most evident and scandalous manifestation of the 'ontological exposure' of the human being, of its inability to find exclusively 'in itself' the resources necessary for its own existence, and hence of its originary dependence on an 'outside', whether nature or society, which determines its position as an 'ex-position': "Human misery is not created by the extreme affliction that falls upon some human beings, it is only revealed by it" [1, p. 262]. Weil affirms that extreme misery *afflicts* human beings. She uses a transitive verb that seems to indicate a clear separation between the complete existence in itself of a human being and the affliction that can come from the outside and 'afflict' it, wound it. However, in the second part of the sentence, she adds that affliction [*malheur*] doesn't 'create' but rather 'reveals' human misery. In reality, however, human misery consists precisely of being ontologically exposed to affliction. Being exposed to affliction 'is' vulnerability, which, as the etymology of the word teaches us, means being exposed to *vulnus*, to 'wounding'. Such an exposure is not contingent but 'internal' to the human being. However, one is rarely entirely conscious of it, except in cases when the intensification of the *malheur* renders our being vulnerable evident, as in the case of a disease or an accident.

In the absence of the specific manifestation of affliction, only a certain degree of 'attention' may result in a 'continuous' awareness of affliction as a condition internal to the human being. It is towards this degree of attention that we must strive, according to Weil: "the soul must be vulnerable to the wounds of all flesh, without any exception, as it is to those of one's own flesh, neither more nor less so; to every death as it is to one's own death" [1, p. 281].[2] In addition, yet, many things lead one to 'evade' such an awareness. This evasion is not always or necessarily the result of contingencies such as the greater or lesser spiritual power of each person individually. The reason this happens is more fundamental, namely, the fact that affliction confronts us with 'life laid bare': "Affliction, under this aspect, is hideous, as life in its nakedness always is; like an amputated limb, or the swarming of insects.

---

[1] One of the reasons why this issue has not been adequately explored is because confronted with the *malheur*, as Simone Weil says, recoiling is the spontaneous response: "Thought revolts from contemplating affliction, to the same degree that living flesh recoils from death. A stag advancing voluntarily step by step to offer itself to the teeth of a pack of hounds is about as probable as an act of attention directed towards a real affliction, which is close at hand, on the part of a mind which is free to avoid it" [2, p. 85].

[2] However, in the *malheur* there is something 'inassumable', as Levinas argued about suffering: "Suffering is, of course, a datum in consciousness, a certain 'psychological content,' similar to the lived experience of color, sound, contact, or any other sensation. However, in this very 'content' it is an in-spite-of consciousness, the unassumable. The unassumable and 'unassumability'" [3, p. 91]. About Levinas' philosophy see [4].

Life without form. Survival is then the one and only attachment [...], without any other object than itself. Hell"[3] [1, p. 223].

Affliction, therefore, can reduce a human being to 'life without form', compelled to look only after its 'persistence'. In this sense, Weil can indeed be considered perhaps the first philosopher to have taken the question of 'bare life' as the specific object of her reflection before Levinas, Arendt or Foucault. The only thinker to have written about it in those same years was Walter Benjamin. "Life without form", "life laid bare": this is the 'supreme vulnerability', the 'zero degree' of the human being, the 'body exposed to wounding and death', ontologically as well as at the hands of others. Although Simone Weil barely knew anything about life and death in camps, her description in her text on the Iliad of the effects of violence on human beings reduced to "cadavers without being dead yet" more than resembles the descriptions we find in the memoirs of Nazi camp survivors:

> this indefinable influence that the presence of another human being has on us is not exercised by men whom a moment of impatience can deprive of life, who can die before even thought has a chance to pass sentence on them. In their presence, people move about as if they were not there; they, on their side, running the risk of being reduced to nothing in a single instant, imitate nothingness in their own persons. Pushed, they fall. Fallen, they lie where they are [...] It is not that their life is harder than other men's nor that they occupy a lower place in the social hierarchy; no, they are another human species, a compromise between a man and a corps. [6, pp. 7–8][4]

To devote attention to human beings reduced to 'bare life' is not only a difficult but almost impossible task, as such attention both confronts and connects us to an *extreme* possibility of the human being, which is still a human possibility to which we are all exposed. This is why Weil takes education to recognise such vulnerability and assume responsibility for it to be among the highest tasks of politics.

Attention has to be educated to identify the *needs* of the body and spirit of each human being to which we are beholden to respond, to be responsible for, before claiming any rights for ourselves. The obligation towards another human being is neither recognised nor acknowledged when the human being is reduced to "bare life". Precisely because of this, Weil recalls that, even if unrecognised, the obligation *exists*, persists and resists: "the object of any obligation, in the realm of human affairs, is always the human being as such. There exists an obligation towards every human being for the sole reason that he or she is a human being, without any other condition requiring to be fulfilled, and even without any recognition of such obligation on the part of the individual concerned" [8, p. 4][5] (Fig. 1).

[3] It is no coincidence that Levinas strongly proclaims the uselessness of all suffering: "Thus, the least one can say about suffering is that, in its own phenomenality, intrinsically, it is useless: 'for nothing.' Doubtless this depth of meaninglessness that the analysis seems to suggest is confirmed by empirical situations of pain, in which pain remains undiluted, so to speak, and isolates itself in consciousness, or absorbs the rest of consciousness" [3, p. 93]. I focused on this issue in [5].

[4] The question of the relationship between subjectivity, power and vulnerability is the core of [7].

[5] We have seen during the Covid 19 pandemic, how the concept of 'obligation to the other' has been used by all governments that have decided to practice policies of restricting freedom and lock

Fig. 1 Early months of the Covid-19 pandemic in Italy. Military trucks carrying coffins that the Italian city of Bergamo can no longer bury, that it can no longer even cremate. The military will escort them to Modena and Bologna, then the ashes will return to their loved 'ones'. Bergamo (Italy)—March 2020

One cannot but begin with this 'ontological' given in order to constitute an ethical discourse and a political project. As important as this invitation of Weil may be, it suffers from one important limitation: namely, it places excessive weight on the subjective disposition in recognising and assuming the vulnerability of others. This is why Weil's philosophical approach can be complemented with that of Judith Butler, helping us rethink the "politics of vulnerability" in a much more systematic way.

3 Differential Vulnerability and Interdependence

What I am particularly interested in here is one aspect of Judith Butler's thought: her attention to all types of vulnerability and all types of 'precarious' life. Like Weil, Butler is famous for not separating subjective questions from broader political reflection. Her philosophy is inserted in the specific line of feminism that insists on the necessary relationship between the personal and political. As noted, "life worth mourning" is one of the modalities Butler uses in trying to give political and philosophical dignity to "precarious life". A play on Levinas, who in his text *Peace*

down. The issue of obligation, in that case, is intertwined with that of community and immunity. To clarify these issues, the work of Roberto Esposito is indispensable: see [9, 10].

Fig. 2 Migrant shipwreck in the Mediterranean Sea in the waters in front of the village of Steccato di Cutro, Crotone province, Italy. More than 72 victims. Crotone (Italy)—March 2023

and Proximity speaks of "the face as the extreme precarity of the other". Butler thinks the precariousness of life in degrees of 'gravity' so to speak, and in ways not dissimilar to Weil, while dedicating specific attention to the multiple forms of contemporary precariousness. Precariousness as the uncertainty of work in the future; as extreme poverty; as lack of primary resources (water, food); as lack of security in war zones; as exposure to gender-based violence, to take only a few heterogenous examples; all the way to "unreal" lives, to the *invisible* lives of those who die of hunger, wars, wreckage, those whose faces and names remain unfamiliar (Fig. 2).

The essential question, then, ought to be: who or what renders such lives precarious or even "unreal"? The unequal distribution of the precarity *differential*, globally as well as within nation-states, is always the outcome of certain orders of discourses and dispositives of power. The latter prevent precariousness from being recognised as such to the point of "normalising" it, presenting it as governed by norms.

Butler develops the concept of *grievable life* reflecting on the events preceding and following *September 11th*: the Americans, after the great *vulnus* inflicted on the nation held to be invulnerable, produce a very precise discourse of power: one has to identify with the victims, recognise their lives as grievable lives, lives worthy of national mourning. Ground Zero, the place where the genuine politics of mourning inseparable from the question of identity was put into place and enacted, demonstrates it clearly. Starting from these premises, Butler asks why some lives are considered worthy of mourning, while others are not. What is at play in this recognition of the life of the other as a grievable life? Such a question, according to Butler, includes or presupposes a certain kind of relationship between the *vulnerability* of the other and one's own. Faced with the attacks, the reaction of the US was that of the total *rejection* of its own vulnerability and, in return, the will to affirm its own power. There was even no 'tension' between invulnerability and vulnerability: vulnerability was externalised, placed outside of itself, celebrating the dead and demonstrating its own power through the killings of others held to be responsible for the *vulnus*. This refusal of one's own vulnerability prevents seeing the vulnerability of others, as accepting one's own vulnerability amounted to *questioning one's own identity*. Admittedly, vulnerability brings into question a *certain* conception of identity, that of a closed, defined, self-centred, self-sufficient subject, confined to both the individual and political level: the sovereign and autonomous subject corresponds to a sovereign state of closed and precisely determined and delimited borders.

Levinas was the first to masterfully *deconstruct* this type of subjectivity, a project continued later in philosophy by Derrida and by Freud and Lacan in psychoanalyses. The latter two revealed that there were multiple elements working within and outside of the I, thus putting an end to any unitary and noncontradictory image of identity. Butler keeps these different lines of thought present, complementing them with feminist psychoanalysis, from Melanie Klein to Julia Kristeva. This is what led Butler to question, for example, *obligatory heterosexuality and the gender binary*. Levinas, however, is certainly the one whose role is most important: the I is always already 'exposed' to the other, not only in the sense that the other is before oneself but that one is lacerated by the other from within, by the Face of the other which commands one not to kill her. Nonetheless, to be vulnerable for Butler—the "thou shall not kill" which, on Levinas' account, the Face of the other commands—isn't a commandment endowed with an intrinsic ethical power capable of modifying relations. Because of this, it can remain ineffectual at the level of both ethics and politics. Vulnerability, for its part, is neither good nor bad per se. It becomes an ethico-political problem once the dispositives of power and dominant order of discourses turn it into a state of precariousness or even "life laid bare":

> So when we say that every infant is surely vulnerable, that is clearly true; but it is true, in part, precisely because our utterance enacts the very recognition of vulnerability and so shows the importance of recognition itself for sustaining vulnerability. We perform the recognition by making the claim, and that is surely a very good ethical reason to make the claim. We make the claim, however, precisely because it is not taken for granted, precisely because it is not, in every instance, honoured. [11, p. 43]

Culture and its norms lead us to agree that an infant is vulnerable. Foucault would say that it is by virtue of the 'discourses' that a certain culture has constructed around infancy that we become capable of recognising the new-born as a vulnerable being in need of care. If, however, we turn our attention to adults, Butler underlines, recognising their vulnerability is not as simple as it might seem. We can give several concrete examples: think of a person with Alzheimer or a person with psychiatric problems. When confronted with these problems, our attention tires and shifts away, not only because of the personal impact but also because of the social "shame" such problems cause due to ingrained prejudices.

However, according to Butler, precise situations such as these can be helpful in illuminating a more fundamental dimension: the vulnerability of such subjects reveals the dimension of the inescapable dependency on others. These adults and their families cannot live a dignified life if not surrounded by networks of assistance and support. In a certain sense, the vulnerability of one is conclusive evidence of the *interdependence* of all, of the fact that being in the world, as children teach us since their birth, is to arrive in this world fundamentally lacking: each and every one of us, even if not ill, depends on the life of others. Interdependence is a constitutive condition of the human, and our vulnerability reveals it as such. Considering all of these elements together, Butler aims to offer an even more precise synthesis:

> The first is that vulnerability cannot be associated exclusively with injurability. All responsiveness to what happens is a function and effect of vulnerability—being open to a history, registering an impression, or having something impressed upon one's understanding. Vulnerability may be a function of openness, that is, of being open to a world that is not fully known or predictable. Part of what a body does (to use the phrase of Deleuze, derived from his reading of Spinoza) is to open onto the body of another, or a set of others, and for this reason bodies are not self-enclosed kinds of entities. [12, p. 149]

What then, is to be done politically? Above all, prioritising and promoting discourses that favour recognition of vulnerability not so much as an 'ontological' element relative to the humanity of the human being– something presupposed, lacking direct ethico-political effects– but as the contingency subject to change in relation to time, place and culture, taking into account its differential distribution as the outcome of certain order of discourses and concrete contingencies. Butler is wary of an 'essentialist' recognition of vulnerability, which, as she underlines, can lead to the emergence of aberrant discourses, such as paternalistic discourses about women or those at the basis of 'humanitarian' wars. Essentialist recognition can ultimately function as an instrument to be used against the discriminated subjects, inverting positions, such as when the 'traditional family' recognises itself as vulnerable in relation to gay families, which then become the enemies to be defended from and fought against.

Instead, Butler is convinced that the relationship between vulnerability and dependency can be politically used in a performative way within certain types of 'nonviolent struggles', which take advantage of being in common in such a way as to expose and bring into crisis oppressive governmental dispositives.

Therefore, to accept the fact that we depend on others is to accept our own vulnerability as cosubstantial with the humanity of the human being. This reveals and points to a more general and complex dependency that has to be understood and grasped so that it can be translated into ethical and political discourse. Vulnerability becomes evident in more or less grave situations, in principle unpredictable and uncontrollable, to which every human being is exposed: from some unpleasant comment of a passer-by, through break-ups of friendships or love relationships, to the annihilation of the entire population by bombing.

This is why I believe it is useful to conclude with one last Butler's quote that explicitly reveals the inextricable relationship between vulnerability, interdependency and political struggle: "To say that any of us are vulnerable beings is to mark our radical dependency not only on others, but on a sustaining and sustainable world" [12, p. 150]. Concretely, this means that when those who are seen as "dispensable" or "not worthy of mourning" unite in public with particular demands, they are enacting a deliberate form of exposure and persistence, "the embodied demand for a liveable life that shows us the simultaneity of being precarious and acting" [12, p. 153].

Taking the thesis I began with, I would say that prior to anything else vulnerability is constitutive of the humanity of the human being; second, that it cannot be separated from a certain order of discourses governed by the powers that be; third, assuming vulnerability as a common trait of humanity is indispensable if we are to commit ourselves to nonviolent coexistence; and finally, envisioning such a coexistence demands extreme attention and public recognition of vulnerability, in particular in circumstances and conditions where vulnerability is not evident.

I think that the steps needed to turn vulnerability into a building block of 'nonviolent' ethics and politics imply that we refuse to consider identity as a self-enclosed monad, as protected and in need of protection. Acceptance, therefore, that identity is always already traversed by the other and open to the outside; moreover, that is always already constituted by the relationship between the outside and the inside. Accepting this fact is a precondition for understanding vulnerability not as a threat but as the condition of being-together-in-one-place, as Levinas' beautiful expression has it. Therefore, there is a demand to rediscover the positive and politically productive meaning and worth of vulnerability, understood as openness to the other and the cause of our mutual interdependency.

If this is a task to be undertaken, above all on a subjective and personal level, the supreme responsibility of a politics that takes vulnerability as its beginning and its end would be that of avoiding at all costs the reduction of any human being to the maximum degree of affliction [*malheur*], to "life without form", exposing it to politically avoidable circumstances. Second, in cases where it is clear that some human beings live in grave conditions of affliction, the responsibility of politics is to remedy these by proposing a solution, having the courage to—radically if needed—question itself to assume responsibility for vulnerability. All institutions should assume this responsibility if they truly believe in justice, which alone is the basis for a nonviolent being-together-in-one-place. This should also be the aim of any instituting process capable of responding to the needs of the present [13, 14].

Core Messages

- Vulnerability is *constitutive* of the humanity of the human.
- Vulnerability also stems from certain discourses of power.
- To accept vulnerability as the common trait of humanity can be the basis upon which to construct a non-violent or least violent possible coexistence.
- In order to arrive at this, the philosophical and political problems to be addressed are *attention for* and *recognition* of vulnerability, and this implies circumstances where vulnerability is not obvious or is not recognised as such.

References

1. Weil S (1956) The notebooks of Simone Weil, vol 1. Putnam's, New York
2. Weil S (2005) Human personality in Id, an anthology. Penguin Books, London
3. Levinas E (1998) Useless suffering, in id, Entre Nous. on thinking-of-the-other. Columbia University Press, New York
4. Fulco R (2013) Essere insieme in un luogo. Etica, politica e diritto nel pensiero di Emmanuel Levinas. Mimesis, Milano-Udine
5. Fulco R (2022) Life and useless suffering: responsibility for others and the impossible theodicy. In Benso S (ed) Rethinking life. Italian philosophy in Precarius Times. SUNY, New York, pp 145–158
6. Weil S (1991) *The* Iliad *or the poem of force*. Pendle Ill, Wallingford Pennsylvania
7. Fulco R (2020) Soggettività e potere. Ontologia della vulnerabilità in Simone Weil. Quodlibet, Macerata
8. Weil S (2002) The need for roots. prelude to a declaration of duties towards mankind. Routledge, London and New York
9. Esposito R (2011) Immunitas. the protection and negation of life. Polity Press, Cambridge
10. Christiaens T, De Cauwer S (2020), The biopolitics of immunity in times of COVID-19: an interview with Roberto Esposito: https://antipodeonline.org/2020/06/16/interview-with-roberto-esposito/?fbclid=IwAR16BcM07RjGaOOpmJkL3qn8JSBn2UsqXx8FB7dVhBK9WL-lP8FSUyuuzz8
11. Butler J (2004) Violence, mourning, politics. In: Butler J, Precarious life. the power of mourning and violence. Verso, London-New York
12. Butler J (2015) Bodily vulnerability, coalitional politics. In: Butler J, Notes toward a performative theory of assembly. Harvard University Press, Cambridge (MA) and London
13. Esposito R (2021) Instituting thought: three paradigms of political ontology. Polity, Cambridge and New York
14. Fulco R (2021) A political ontology for Europe: Roberto Esposito's instituent paradigm. Cont Philos Rev 54:367–386. https://doi.org/10.1007/s11007-021-09542-z

Rita Fulco is Assistant Professor of Theoretical Philosophy (University of Messina, Italy).

She was postdoc fellow in Theoretical Philosophy, from 2016 to 2021, at the Scuola Normale Superiore, Pisa (Italy).

She has worked on the theoretical, religious, ethical and political entailments of twenty-first century's continental philosophy. Among her publications: *Corrispondere al limite. Simone Weil: il pensiero e la luce* (Studium 2002); *Il tempo della fine. L'Apocalittica messianica di Sergio Quinzio* (Diabasis 2007); *Essere insieme in un luogo. Etica, politica, diritto nel pensiero di Emmanuel Levinas* (Mimesis 2013); *Soggettività e potere. Ontologia della vulnerabilità in Simone Weil* (Quodlibet 2020); R. Fulco, T. Greco (eds), *L'Europa di Simone Weil. Filosofia e nuove istituzioni* (Quodlibet 2019); R. Fulco, A. Moresco (eds.), *Sull'evento. Filosofia, storia, biopolitica. Almanacco di Filosofia e Politica 4* (Quodlibet 2022); *Life and Useless Suffering: Responsability for Others and the Impossible Theodicy*, in S. Benso (ed.), *Rethinking Life* (SUNY 2022).

Chapter 4
Vulnerability and the End of the World. Trying to Read the Post-pandemic Age (with Karl Jaspers and Ernesto De Martino)

Anna Donise

In the experience of the end, what ends is first and foremost
the signifier, what is operable according to values,
the intersubjective and communicable communal design,
the power to always go beyond the situation
Ernesto De Martino, La fine del mondo, 2002.

Abstract This essay aims to explore the notion that the preventive measures and restrictions imposed during the pandemic exposed a significant aspect of our vulnerability. The pressure to maintain distance from objects and people and recognize their potential threat compelled us to adopt the behaviors of those afflicted with delusions, such as end-of-the-world hallucinations. With the virus being invisible, we have been required to act as though it was always present, in a way akin to a collective delusion. Consequently, this shift in our everyday actions has caused collective psychological distress and evident existential anguish: mimicking madness leads to pathological experiences. While recognizing the importance of preventive measures, this paper argues for the necessity of raising cultural awareness to comprehend the source of these distressing experiences and incorporate them into a more comprehensive concept of community. In conclusion, the paper advocates for the acquisition of cultural understanding to enhance our ability to respond to crises in a way that considers the well-being of all members of our community.

Keywords Crisis · Psychopathology · De Martino · Jaspers · Well-being

---

A. Donise (✉)
University of Naples "Federico II", Naples, Italy
e-mail: anna.donise@unina.it

S. Achella and C. Marazia (eds.), *Vulnerabilities*, Integrated Science 18,
https://doi.org/10.1007/978-3-031-39378-5_4

1 Introduction

Many intellectuals have been called upon to reflect on the relationship between the pandemic and our society, investigating the theoretical and social consequences of policies implemented by different governments. Italian thinker Giorgio Agamben exemplifies those who have taken explicitly denialist positions,[1] later even embracing hints of conspiracy theory.[2] The position of French philosopher Bernard-Henri Lévy is slightly different and more relevant to the argument presented in this paper. Like Agamben and many others, in his 2020 pamphlet [2], Lévy argues that the COVID-19 pandemic was not fundamentally different from many recent experiences in terms of its actual effects.[3] The real difference, he suggests, lies in our perception of the event: "the entire planet—rich and poor alike, those with the resources to resist and those without—pounced on the idea of an unprecedented pandemic poised to eradicate the human race" [2, pp. 12–13]. The truly unique quality of these pandemic years stems from the sense of impending apocalypse, the feeling of an end-of-the-world experience that went with them. This feeling arose from the constant information bombardment by the media and social networks, continuously reporting on the number of deaths and the situation in overcrowded intensive care units: those outlets "herded us into a parallel universe in which nothing else, anywhere, was news—and, in so doing, drove us quite literally mad" [2, p. 14]. Hence the suggestive title of his pamphlet: *The Virus in the Age of Madness*, which in the original French version is, in my opinion, even more interesting—*Ce virus qui rend fou.*

Yet the philosopher's response to the connection between the virus, apocalyptic experiences, and madness is ultimately underwhelming. Indeed, Lévy's position aligns with others who oscillate between anger and denial, blaming the media and governments for the collective state of mind that has characterized different countries as well as varied cultural and political contexts. While communication undoubtedly played a significant role in the reactions and experiences of the community during these challenging pandemic years, philosophical analysis can delve deeper and attempt to interpret other elements to support our understanding of the crisis. So let us begin with the intriguing connection suggested by Lévy between the first pandemic experienced in a globalized and interconnected world and the sense of approaching the end of the world, which he rightly considers an experience *qui rend fou.* However, the philosopher only briefly touches on this connection, missing an opportunity, I think, to offer tools to better comprehend our experiences during the pandemic and even beyond.

---

[1] The views expressed by Giorgio Agamben commenting on the pandemic can be found on the website of the publisher Quodlibet, where he writes the column "Una voce. Rubrica di Giorgio Agamben". His first reflection devoted to COVID-19 is titled "Invenzione di una pandemia" (Invention of a Pandemic) and is dated February 26, 2020.

[2] For a comprehensive reading of the very extensive literature on the subject, see Maurizio Ferraris' contribution [1].

[3] Lévy refers not only to the Spanish flu, which caused 50 million deaths, but also to the Hong Kong flu of 1968 and another flu, also of Asian origin, due to which "as many as two million perished" [2, p. vii ff.].

The following pages propose a reading of the pandemic, starting with an investigation of "end-of-the-world experiences" to highlight similarities with the pandemic situation. My thesis is that the pandemic compelled us to replicate behaviors typical of delirium, transforming our daily actions and relationships with the world. This transformation has caused much distress and anxiety, making us feel "mad", as it were, and confronting us with our structural vulnerability. This analysis highlights the importance of the ways in which we operate and interact with the world in constructing our identity and well-being. It also provides a lens through which to view vulnerability, seeing it as a fundamental aspect of our existence that we compensate for by building relationships and social structures that function as protective strategies in times of crisis. The pandemic years have made it clear that we cannot do without such coping strategies, which serve as shells and rituals of protection in the face of our structural vulnerability.

2 End-of-The-World Experiences

The strategy of comparing psychic pathology with social pathology is not a new approach. However, the interesting aspect, in this case, is the type of delirium that Lévy refers to, namely, apocalyptic delirium. Therefore, it is worth describing in more detail the experiences that are characteristic of so-called "end-of-the-world delusions," a psychopathological condition that has a corresponding communal and social level: the impending sense that a crisis of an era, a cultural and value crisis, will culminate in an authentic cultural apocalypse.

The experience of delirium has been subject to countless attempts at definition, but each one seems to capture only one aspect [3, p. 61]. The awareness of the deep interrelationship between bodily, social, cultural, and psychological dimensions appears to progressively limit reductionistic interpretations, even in medical and scientific contexts [4]. End-of-the-world delirium, which is the subject of interest here, is often preceded by a strong feeling of uncertainty. Patients feel a sense of alienation and confusion, and all communication with the world becomes unsure and problematic [5]. Karl Jaspers, the German psychopathologist and philosopher who initiated the era of phenomenological psychopathology,[4] described this phase very well through the example of a patient at Sandberg who kept telling her husband: "something is going on; do tell me what on earth is going on" [9, p. 98]. Her disorder did not alter her perception of the world, which remained unaffected. Rather, she felt a subtle change in everything, giving her an uncanny feeling (*Unheimlichkeit*) [9, p. 99]. More generally, at this early stage of delirium, things seem to take on different meanings, and the whole world appears to be altered. This change is not sensory; in

[4] In 1912, Jaspers wrote an essay entitled *Die phänomenologische Forschungsrichtung in der Psychopathologie* (The Phenomenological Research Direction in Psychopathology), which gave rise to phenomenologically oriented psychology. On this topic, cf. [6–8].

contrast, perceptions remain unchanged. Rather, there is a subtle shift that pervades everything and creates an unsettling, bewildering atmosphere.

However, it is important to distinguish between a properly delusional perception, which immediately grasps the delusional element as already structured and defined, and delirium without an object—a "delusional atmosphere (*Wahnstimmung*)" [9, p. 98], or mood, in which the sufferer senses something disturbing and sinister in the environment without being able to identify it, as in the case of the Sandberg patient. The delusional mood is particularly unnerving, and when an object can be identified as the origin of the mood, this is experienced with great relief. In the case of end-of-the-world delirium, the delusional mood remains constantly present and takes the form of a persistent sense of change, accompanied by an attempt to defend oneself against something dangerous and distressing. To provide an example drawn from Wetzel [10], a 38-year-old worker stated that to her everything was equivocal, as if everything in the world were radically different (*ganz anders*). While objects present themselves as meaningful and embedded in a world that appears to make sense for a "normal" individual, for someone with end-of-the-world delusions, the objects of everyday life have an entirely different meaning.

Italian anthropologist Ernesto De Martino (1908–1965) devoted particular attention to the analysis and possible correlation between psychopathological and cultural apocalypses, both in the unfinished work *La fine del mondo* (The End of the World) [11] and in his *Apocalissi culturali e apocalissi psicopatologiche* (Cultural Apocalypses and Psychopathological Apocalypses, 1964) [12]. In end-of-the-world delirium, he argues, the objects of everyday life completely change their meaning: "under certain conditions [they are] the objects of everyday interaction, under certain other respects they manifest a whole other, demonic essence" [11, p. 43]. In the first case, things appear no longer adequate to their functions; they become insignificant, and they do not prompt any operation, leaving the subject exposed and vulnerable. But even more interesting is the second case, that of "*too much*" semanticity. In that case, objects appear to be enveloped in "a dark semantic halo" and even exceed "their domestic and obvious borders to become indices of an indeterminate beyond" [11, p. 87]. Objects do not convey connections, but rather distance, a constant sense of being out of place. They seem to allude to something that is clearly no longer their domestic function, so "every perceived object, because of this futile search for its beyond, is in tension, and its dark semantic halo is experienced as catastrophic anticipation. The universe is a universe in tension" [11, pp. 88–89]. When things cease to be understandable and refer to unforeseen and unpredictable meanings, subjectivity loses all forms of defense and experiences its own vulnerability as an unbearable condition.

However, it is not just the world of objects that changes for the subject experiencing end-of-the-world delusion, but also the relationship with other subjects. This raises the question of what it means for the sufferer of this kind of delirium to lose their "being-in-the-world".[5] The sufferer can no longer maintain a relationship with others and finds themselves "thrown out of being-with," removed from their "*Miteinander sein*" [14]. The individual feels that they are no longer a part of the common world and are "placed in a private world" [15]. This element further emphasizes the dramatic nature of apocalyptic experiences: humanity experiences the loss of intersubjective relations, which brings out a shared world of values—a world of values that make the human world possible. Hence a sketch of the typical human condition, that which constitutes fundamental character of its "normality," that is, the possibility of acting with others: "its designable intersubjectivity, its belonging to a socially and culturally conditioned perspective of operability" [11, p. 50]. To lose the normality of the world is to lose the ability to communicate one's private world by turning it into a word or gesture placed "in a dynamic of intersubjective valorization."

Indeed, "one looks with suspicion or dismay or pity at those who spend their days magnifying the ineffable they carry within, the treasure they hide in their chests" [11, p. 50]. The dimension of intersubjective doing and acting becomes the place where one can "make oneself healthy." The inner dimension, on the contrary, is healthy only when it is at least implicitly oriented toward communication and intersubjectivity. After all, the end-of-the-world experience is nothing but the crisis of this intersubjective project, the crumbling of "the ethos of valorizing transcendence in the inaugural value of the communal project of the usable" [11, p. 77]. From this perspective, the real difference between "normal" and "abnormal" lies in the absence of valorizing momentum that transcends the dimension of life for those who experience forms of delirium. This ethos of transcendence is what makes us participants in an intersubjective world.[6] It can be argued that psychic abnormality is not the inability to follow a norm, but the inability to be normative and produce shared cultural forms. As De Martino puts it, "the fall of this momentum—whatever the hereditary or acquired somatic events when one considers such a fall from a medical-operational perspective—is (…) the receding of the power of transcendence on the whole front of the valorizable, the catastrophe of the valorizing momentum" [11, pp. 15–16].

3 Responses to the Crisis

As discussed thus far, De Martino sees the psychopathological apocalypse as the ultimate risk, an irredeemable collapse. However, the apocalyptic dynamic, connected to the cultural horizon, is investigated from an entirely different perspective. The thesis is clear: apocalypse and apocalyptic experiences are a constant anthropological dynamic, and the only possible response and reaction to the crisis they reveal

[5] De Martino's analysis is built in dialogue with authors who drew on analytic-existential psychiatry. In particular, he cites Alfred Storch, Caspar Kulenkampff, Hans Kunz, as well as van den Berg of the Utrecht School, an author who advocated a phenomenological-existential approach [13].

[6] The idea that the breakdown of intersubjectivity underlies psychic pathology is also very much present in contemporary psychiatric literature [16–18].

is culture, interpreted as a shared human activity. The comparison with psychic pathology serves a functional purpose, and the use of the heuristic potential of psychopathological experiences enables "the moment of risk to be exposed with particular clarity" [11, p. 18], bringing out the crisis element that healthy individuals overcome and thus "cover up." Research on cultural apocalypses employs "the morbid to shed light on the process of 'making oneself healthy' that characterizes culture, at least as long as it manages to function" [11, p. 19].

The concept of cultural apocalypse can be concisely defined as the "manifestation of cultural life that involves, within the framework of a particular culture and historical context, the theme of the end of the current world" [12, p. 105]. Every culture contains the idea of periodic or potential destruction and rebirth, and cultural apocalypses take various forms, ranging from archaic depictions of the eternal return to the Christian apocalypse, which progresses from the prediction of an imminent end of the world to the announcement of the Kingdom of God. Even the Marxist notion of the end of a world (the bourgeois one), followed by the advent of a new Proletarian world, can be considered apocalyptic.

The cultural apocalypse, while indicating a risk, also contains within it the potential for rebirth precisely because it is embedded in a cultural context. In fact, many forms of defense and reaction against psychic disintegration operate through shared communal actions.[7] A good example is that of mourning, which Jaspers had already defined as a "borderline situation" [19, p. 201]. If it becomes an illness, "it is not susceptible to any cultural history," but culture as a whole provides "the resources to go beyond the critical moment of the mournful event and to overcome (…) the temptation of crisis." It is only in this "negative moment of cultural redemption" that mourning as an illness "enters history" [21, p. 53].

Cultural and historical life has developed various techniques to facilitate mourning, and one such technique is the funeral lament, to which De Martino devotes extensive space. According to him, funeral lament is a system "of institutional dehistorification of death". At critical moments of existence, when becoming creates anguish, rituals—particularly religious ones—manage to accomplish a kind of "erasure or masking of the distressing history," giving rise to "dehistorification" [22, p. 62]. These techniques involve the repetition of a refrain in lament or other rituals, through which the infinite historical variety of mournful situations "is dehistorified into mimetic, melodic, and literary patterns that are fixed in the cultural memory of communities and repeated as a ritual obligation" [21, p. 310]. By acting, even through mimed or repeated actions, the possibility of overcoming crisis can be reopened (see Fig. 1).

It is therefore important to distinguish between what can be understood as "the crisis of a single valorization," that is, going through a critical experience or a difficult situation, on the one hand, and what we can call "the fall of the ethos of transcendence

---

[7] Psychic pathology itself can be read as a last, desperate bulwark of acting, by putting up the most strenuous resistance: just think of catatonic forms as "strongholds of permanence" erected by humans even on the now crumbling ground of pathology. On the concept of the "shell" as a coping strategy, De Martino's reference is Jaspers' *Psychologie der Weltanschauungen* (Psychology of Worldviews), published in 1919 [19]. Allow me to refer to my own contribution on the topic [20].

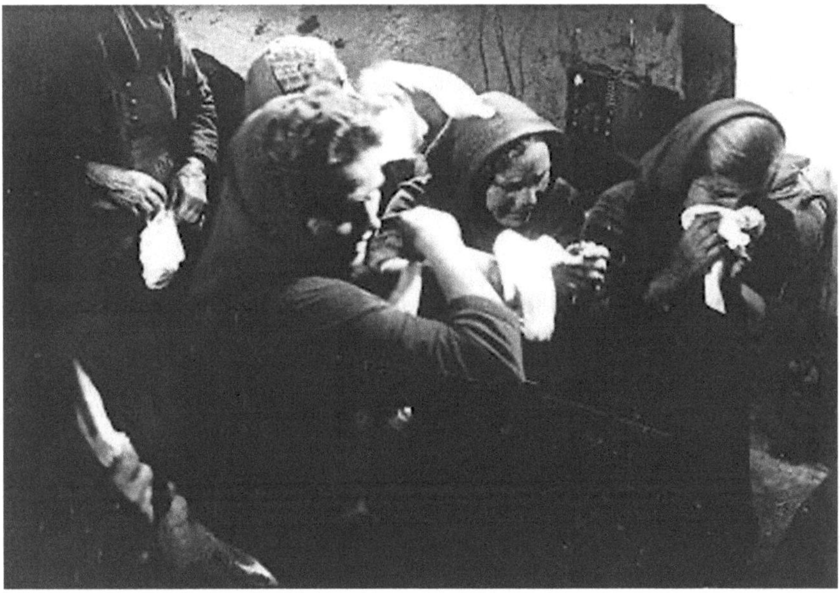

Fig. 1 Morte e pianto rituale (Death and ritual weeping)

(the risk of not being able to exist in any possible cultural world)" [23, p. 7], which characterizes the ultimate condition of psychiatric illness: dementia and death.[8] After all, Jaspers had already argued that borderline situations "do not offer a fixed point, an indubitable absolute element, a support that gives firmness and stability to every experience and every thought. Everything flows […] everything is relative, split into contraries" [19, p. 202]. Borderline situations, such as death and mourning, are unbearable for the individual experiencing them. It is not surprising that in daily life, they are almost never experienced to their full extent. Instead, it is far more common for humans to find some foothold in the face of collapse. Without the possibility of locating a foothold, life itself would cease.

The mythic-ritual symbolism that De Martino speaks of can be understood as one of the possible "footholds" mentioned by Jaspers. Ritual acting is entrusted with a crucial task, which is to protect humans from the disruptions of historicity. In brief, the idea is that the crisis, which is a permanent anthropological risk, can be overcome through a series of strategies centered on shared action. Psychic pathology is

[8] It is not surprising, then, that De Martino states that the themes of Paci's and especially Abbagnano's positive existentialism "largely fit with the perspective I chose for the monograph on the 'end of the world'" [23, p. 71]. Nicola Abbagnano views existence as transcendence, as the possibility of a relationship with being. The fundamental structure of humanity is the tension in the realization of an ought-to-be. The first principle is not being but the possibility of being, an ought-to-be that must be realized, which becomes a task given to individual existence. By recognizing the relevance of human possibility, Abbagnano redeems "the world of the usable," which Heidegger relegated to "inauthentic existence" [24, pp. 34-35], in De Martino's perspective.

disintegrative: it is the risk against which rituals and mythic-ritual symbolism defend society. Rituals, which involve mimed and repeated actions, serve as a normalization mechanism in the face of the risk of loss.

4 How to Become Mad

Let me now summarize the salient points that have emerged thus far, in order to establish a comparison with the pandemic situation and argue my thesis. In end-of-the-world experiences, human beings are exposed and vulnerable in that:

1. There is a shift in the semantic framework of objects, which lose their usual operability and acquire obscure meanings that refer to danger and some imminent catastrophe.
2. Other subjects become imbued with "abnormal" meanings and are no longer part of our common world but instead represent danger.
3. Intersubjective and communal practices, such as rituals, become important coping strategies in the crisis.

I will expand on this last point (3) in the conclusion of this paper. For now, my analysis will turn to the particular kind of vulnerability that emerged in everyday life during the pandemic. If we look at how we lived during those times, through the lens of these reflections on end-of-the-world delusions, we cannot fail to notice common elements. The virus is invisible, making it impossible to know whether it is present in people or objects around us, causing everything to potentially be dangerous to ourselves and others, especially those who are more vulnerable. This required us to act as though everything and everyone around us carried a threat, with meanings different from their usual semantic framework. In essence, we were required to imitate the behavior of delusional individuals, as the danger was not immediately apparent or certain. We had to distance ourselves from others and things, fear them, and be afraid of them, which essentially meant "mimicking and repeating" the forms of end-of-the-world delusion.

The most common objects, from park benches to streetcar seats, from cans at the supermarket to cups at cafes, took on new, disturbing meanings precisely because of this. Anything we touched could be a source of contagion, and we were urged to either avoid contact or immediately sanitize our hands whenever they were exposed to the world and potentially the virus. We were continually asked to change our spontaneous attitude and recognize that these commonplace objects had a new aura that would not connect us but distance us. Their domestic and familiar function was no longer in the foreground, as De Martino's description of end-of-the-world delusions started to ring true for them as well: "every perceived object, because of this futile search for its beyond, is in tension, and its dark semantic halo is experienced as catastrophic anticipation. The universe is a universe in tension" [11, pp. 88–89].

An episode narrated by Sartre, in which he describes what happens to the protagonist of *Nausea* on a tram, could be likened to the experience many had in the

Fig. 2 People waiting to enter the supermarket (Prato, Italy)

days immediately following the lockdown. Upon boarding the streetcar, Antoine Roquentin notices that everything around him has undergone a transformation: "I lean my hand on the seat but pull it back hurriedly" [25, p. 125], he writes. What he describes is the crisis of usability in everyday objects; things lose their connotations of familiarity and take on a life of their own. "The conductor blocks my path. (…) But I push him aside and jump out of the tramway. I couldn't stand any more. I could no longer stand things being so close" [25, p. 126]. The proximity of objects and other human beings has become a source of fear, causing a real sense of panic.[9] Roquentin wanders the streets of the neighborhood, and everything around him no longer appears natural.

The situation becomes even more alarming when it comes to other people. The other subject is perceived as a significant source of risk, and the possibility of contagion arises precisely from their embodied presence, as Husserl would put it. They breathe, speak, move, touch, and even cough—all of which contributes to the danger perceived. The only way to protect ourselves from the other, who scares us, is to maintain distance (as evidenced by the orderly lines outside supermarkets during lockdowns, see Fig. 2) and to use protective devices such as masks or visors to shield ourselves from the other's body.

So, what we had to do, or rather—to emphasize the words De Martino used to describe rituals—what we had to "mimic and repeat" was the act of exiting from

[9] "[…] the houses watched my flight with their mournful eyes. I repeated with anguish: Where shall I go? […] my eyes went rapidly from one (object) to the other, to catch them unawares, stop them in the midst of their metamorphosis. […] Doors of houses frightened me especially. I was afraid they would open of themselves" [25, p. 78].

the intersubjective relation. Similar to what happens to a person with end-of-the-world delusions, we had to feel "thrown out of being-with," taken away from the *Miteinander sein*, the common world, and "placed in a private world" [11, p. 49],[10] which increasingly became our home, experienced as a defensive fortress.

Given the importance of the dimension of communal action, highlighted by De Martino and much psychiatric literature, it is clear that this "living as if" we were suffering from end-of-the-world delusions can have significant effects on our overall psychological balance and well-being. In fact, I believe that this perspective can help us better understand angry and aggressive reactions on the one hand, and depressive and distressed reactions on the other. If the necessary distancing actions required us to assume a horizon similar to "end-of-the-world delusion," then it is not surprising that in some cases, we also adopted the attitudes and ideas that accompany this type of apocalyptic delusion, such as the constant feeling of being the object of plots, "machinations, curses, and the like" [11, p. 17]. From such perspectives, events are designed, and "nothing is random but directed towards the sick person" [11, p. 23]. The lost common horizon is regained through a paranoid interpretation that identifies a cohesive and intelligent community of others hatching plots from which the subject is excluded. Due to space constraints, this cannot be the place to investigate the effects of the collective condition's upheaval on individuals. However, it is sufficient to consider the impact of this transformation of intersubjective relations on developing individuals, such as children and adolescents [26, 27].[11]

5 How to "Make Yourself Healthy"

By imitating delusional behavior, we bring out our vulnerability and become unable to seek refuge in communal relationships. This realization is crucial to fully comprehend the features of post-pandemic suffering. In times of distress, particularly in what Jaspers called "borderline situations," where the individual feels a complete absence of familiar points of reference due to a scary and unforeseeable situation (as during the pandemic's most challenging moments), there is an increased urge to rely on the habitual dimension of mundane daily life. In fact, such a dimension is fundamental to subjectivity's effective action on the world, as only the obvious and customary world provides footholds and "handles" [28] to guide its intentional acts. Our habits, relationships, and customary places offer reassurance and make us feel enveloped and protected. We construct habits and relationships for ourselves that become similar to shells [19, p. 284 ff.], providing protection and containment for our subjectivity in its encounter with a world that is now difficult or even hostile. The return to a reassuring shell is one of the possibilities available to subjects to cope with moments of crisis.

[10] Here De Martino refers to Hans Kunz [15].

[11] Clearly, the topic is very complex and broad. See the many articles that have come out in widely circulated journals such as *The Lancet* or even the *Journal of Youth Adolescence*.

A final consideration is necessary. While our actions during the pandemic certainly resembled end-of-the-world delusions, we must also recognize that what we experienced could and should be seen as a collective effort. If properly directed and narrated, this idea can restore to each individual the awareness that authentic "communal action" during the pandemic necessitated a joint effort to prevent contagion. It is no coincidence that the moments when the difficulty was best addressed were those in which the emphasis was on collective effort and sharing. Being part of a community and recognizing its value then meant being able to sacrifice one's own intersubjective "momentum" and relationships in the name of a broader idea of community that also includes the fragile people who were most directly at risk. In this sense, we witnessed the return of the process that De Martino described as "making oneself healthy" [11, p. 19]: cultural awareness allows one to place one's frustrating and distressing experience within a more complex idea of community, thus recovering one's operational horizon.

Core Messages

- In psychic pathology, end-of-the-world experiences expose the subject to the loss of an operable and usable world, which instead acquires allusive and obscure meanings hinting at danger and impending catastrophe; other subjects also become laden with "abnormal" meanings and are no longer part of our common world but represent a threat.
- If the correct response to the psychopathological crisis is to seek community and shared coping strategies, based on the indications of phenomenological psychopathology and De Martino's anthropological analyses, during COVID-19, we faced the opposite challenge: we were asked to artificially adopt delusional and distancing behavior that led to the fracture of the shared world. The pandemic forced us to mimic and repeat the behavior of a subject affected by end-of-the-world experiences.
- The loss of our ability to operate in the world with others highlights our vulnerability and the importance of relying on our "shells," such as our habits and relationships, to "make ourselves healthy" on our existential journey.
- Furthermore, it is important to recognize that authentic "communal action" required sacrificing our interpersonal momentum and relationships for the sake of the fragile individuals who are most susceptible to the virus.

References

1. Ferraris M (2021) Post-Coronial Studies. Seicento sfumature di virus. Einaudi, Torino
2. Lévy BH (2020) The virus in the age of madness. Yale University Press, London
3. Borgna E (2009) Come se finisse il mondo. Feltrinelli, Milano
4. Wilson JE, Mart MF, Cunningham C et al (2020) Delirium. Nat Rev Dis Primers 6(1):90. https://doi.org/10.1038/s41572-020-00223-4

5. Burkhardt H (1964) Die Wahnstimmung als pathologisches Kommunikationsphänomen. Nervenarzt 35:405–412
6. Donise A (2008) Jaspers als Phänomenologe. Studia philosophica 67:339–352
7. Stanghellini G, Fuchs T (eds) (2013) One century of Karl Jaspers' general psychopathology. Oxford University Press, Oxford
8. Achella S, Donise A, Wagner S (eds) (2014) Psicopatologia generale 1913–2013. Studi Jaspersiani II
9. Jaspers K (1963) General psychopathology. University of Chicago Press, Chicago
10. Wetzel A (1922) Das Weltuntergangserlebnis in der Schizophrenie. Zeitschrift für die gesamte Neurologie und Psychiatrie 78:403–428. https://doi.org/10.1007/BF02867625
11. De Martino E (2002) La fine del mondo: Contributo all'analisi delle apocalissi culturali. Einaudi, Torino
12. De Martino E (1964) Apocalissi culturali e apocalissi psicopatologiche. Nuovi argomenti 69–71:105–141
13. Berg JH van der (1955) The phenomenological approach to psychiatry. an introduction to recent phenomenological psychopathology. Thomas, Springfield
14. Heidegger M (1927) Sein und Zeit. Max Niemeyer, Halle
15. Kunz H (1930) Die Grenze der psychopathologischen Wahninterpretation. Zeitschrift für die gesamte Neurologie und Psychiatrie 135:671–715
16. Stanghellini G, Broome M, Fernandez AV, Fusar-Poli P, Raballo A, Rosfort R (eds) (2019) The Oxford handbook of phenomenological psychopathology. Oxford University Press, Oxford
17. Zahavi D (2014) Self and other: exploring subjectivity, empathy, and shame. Oxford University Press, Oxford
18. Bizzari V (2018) Sento quindi sono: Fenomenologia e *Leib* nel dibattito contemporaneo. Mimesis, Milano
19. Jaspers K (2019) Psychologie der Weltanschauungen: Karl Jaspers Gesamtausgabe Band I/6. Schwabe Verlag, Basel
20. Donise A (2017) Gli involucri tra forma e vita: per una difesa dell'inautentico. Discipline Filosofiche 27(1):175–188
21. De Martino E (2011) Morte e pianto rituale nel mondo antico: Dal lamento funebre al pianto di Maria (1958). Bollati Boringhieri, Milano
22. De Martino E (1995) Fenomenologia religiosa e storicismo assoluto (1953–1954), in De Martino E Storia e metastoria. I fondamenti di una teoria del sacro, ed by Massenzio M. Argo, Lecce 1995:47–74
23. De Martino E (2005) Scritti filosofici. Il Mulino, Bologna
24. Abbagnano N (1948) Esistenzialismo positivo. Taylor, Torino
25. Sartre JP (1964) Nausea. New Directions, Norfolk
26. Creswell C, Shum A, Pearcey S, Skripkauskaite S, Patalay P, Waite P (2021) Young people's mental health during the COVID-19 pandemic. Lancet Child Adolesc Health 5(8):535–537. https://doi.org/10.1016/S2352-4642(21)00177-2
27. Brooks SK, Webster RK, Smith LE et al (2020) The psychological impact of quarantine and how to reduce it: rapid review of the evidence. The Lancet 395(10227):912–920. https://doi.org/10.1016/S0140-6736(20)30460-8
28. Stanghellini G, Ciglia R (2008) The subject of humanities in the light of Ernesto De Martino's philosophical anthropology. Journal für Philosophie und Psychiatrie 1(1), http://www.jfpp.org/jfpp-1-2008-03.html

Anna Donise is Full Professor of Moral Philosophy (University of Naples Federico II). Her research focuses on neokantianism, phenomenology, the concept of "value" and that of "empathy". She is presently working on the theme of the crisis of subjectivity, investigated both on a theoretical level (Husserl, Rickert, Dostoevsky and Mann) and on the level of individual and social pathology (Jaspers, Sartre, De Martino).

She is the author of many volumes and essays, and has translated into Italian several works by Heinrich Rickert, Edmund Husserl and Karl Jaspers. Her recent publications include the book *Critica della ragione empatica. Fenomenologia dell'altruismo e della crudeltà* (2020) and the essays: *The Levels of Empathy. A Phenomenological Contribution to Psychopathology*, (2020); *Empathy and Recognition* (2021); *Moral Dilemmas and Rationality* (2022).

Chapter 5
A Biosocial Perspective on (COVID-19) Pandemic Outbreaks: Interfaces of Biology and Social Determinants

Nico Dragano

Embodiment [is] a concept that refers to how we literally incorporate, biologically, the material and social world in which we live, from in utero to death; a corollary is that no aspect of our biology can be understood absent knowledge of history and individual and societal ways of living
Nancy Krieger, Proximal, distal, and the politics of causation, 2008

Abstract This chapter deals on a theoretical level with the importance of interactions between biological and social factors in the epidemic or pandemic spread of infectious diseases. Biosocial interactions are a fundamental factor in the spread of infections, and a more detailed understanding of these interactions is of scientific and practical interest. To be able to describe systematics, it is proposed to link a classic model of infectious disease epidemiology (the epidemiological triad) and a model of the social determinants of health (ecosocial theory). To illustrate the approach, principles of the biology of infectious diseases are examined more closely to identify possible interfaces with social conditions that make it easier or more difficult for the pathogen to spread. This presentation follows the biological process of an infection from exposure to infection and subsequent disease. Step-by-step, it addresses pathogen characteristics and their implications for the interaction with hosts and their social contexts.

Keywords Social determinants · Infectious disease · Biosocial · COVID-19 · Pandemic preparedness · Ecosocial theory

---

N. Dragano (✉)
Institute of Medical Sociology, Centre for Health and Society, Medical Faculty and University Hospital, University of Düsseldorf, Düsseldorf, Germany
e-mail: Dragano@med.uni-duesseldorf.de

S. Achella and C. Marazia (eds.), *Vulnerabilities*, Integrated Science 18,
https://doi.org/10.1007/978-3-031-39378-5_5

1 Introduction

A pathogen needs the cooperation of its (potential) hosts for an epidemic or even a pandemic to develop. These must either meet in large numbers and become infected in the process or, in the case of indirect transmission, seek contact with the contaminated material in large numbers. This is also the case with infectious diseases that affect humans, which are the focus of this article. It follows that the behaviour of people is a decisive factor in the dynamics of the spread of infectious diseases. However, behaviour is to be understood here in the sociological sense, i.e., not as a purely individual action but as collective, socioculturally shaped patterns of action of larger social groups or entire societies. Of course, a single person who drinks contaminated water, for example, can become infected with cholera. However, a major outbreak will only occur if a large number of people drink from this source and no measures are taken by the group to close or avoid this source. The interactions between pathogens and hosts are, naturally, more complex in reality and depend on both the biological properties of the respective pathogen and the concrete behaviours or reactions of the population affected [1]. A few examples will illustrate this. The transmission properties of a pathogen determine, for example, which social practice increases the probability of infection in the first place. Regarding a respiratory pathogen that is transmitted from person to person via aerosols, different social interactions pose a risk compared with a sexually transmitted disease. Conversely, people at risk of infection react to the characteristics of the pathogen, for example, by deliberately refraining from certain behaviours or by taking organised measures at the social level to specifically combat pathogens or prevent potentially dangerous social situations.

Despite the obvious need to consider pathogen characteristics AND the logic of social practice and social structures when assessing outbreaks, research explicitly addressing the principles of this linkage is comparatively rare [1, 2]. This is understandable, as research in the field of infectious diseases is also primarily disciplinary. Microbiologists, virologists, infectiologists, psychologists, medical sociologists or public health researchers each have their own perspectives on the phenomenon, and even if the contributions of the other disciplines are certainly noticed, genuine interdisciplinary cooperation has been unusual thus far. In addition, the social dimension of the spread of infectious diseases has generally been somewhat out of the focus of research until recently. The social determinants of health (understood as social influences on population health) have been intensively researched, but mainly in relation to noncommunicable diseases, such as diabetes, obesity or cardiovascular disease. Infectious diseases, however, have special characteristics that make it difficult to generalise scientific findings from research on other diseases. Research on social determinants also tends to focus on the social sphere as a source of (causal) influences on health, while the biological processes mediating between social factors and the emergence of a specific disease often play only a secondary role [3]. Of course, there are exceptions in the field of infectious diseases, such as research on the social conditions of the spread of HIV and AIDS, which has revealed, for example, the importance of stigma, the influence of gender norms or the fatal consequences of poverty for diagnosis and treatment [4]. However, little scientific evidence exists for many other infectious diseases, and this applies not only to neglected tropical diseases but also to globally occurring diseases, such as influenza.

Nevertheless, research has made important progress during the coronavirus disease 2019 (COVID-19) pandemic. Within a short period of time, many new empirical studies on different social influences on the course of the pandemic were published [e.g.: 5, 6]. Thus, correlations that were already known could be replicated for COVID-19, and those previously unknown were uncovered. Similarly, there have been initial attempts in connection with COVID-19 to develop further at the theoretical level [e.g. 7]. In addition to new research, the COVID-19 pandemic has also provided ample illustrations of how important it is in the practical management of outbreaks that biological and social processes and their interactions are known to be able to respond in a targeted manner with public health measures. One example is the relatively long-standing uncertainty about the transmission routes (e.g., droplets, airborne, fomite) of the severe acute respiratory syndrome coronavirus 2 (SARS-CoV-2) wild-type and its later variants. This lack of knowledge about the specific sites and circumstances of infection has made it difficult to enact appropriate nonpharmaceutical interventions and protect vulnerable populations at particularly high risk. A lack of knowledge about relevant biosocial contexts thus carried the risk of misdirection (by taking both noneffective or effective measures), which may have had negative consequences for the control of the pandemic as a whole. In response to this experience, there are calls from the scientific community and public health practitioners for a science-based approach to both the biological and social determinants of the COVID-19 outbreak (and future pandemics) to become part of organised global and national pandemic preparedness and response [8].

What exactly such pandemic preparedness should look like regarding biosocial interactions is not easy to answer at present. This is because the scientific foundation is incomplete, as explained above. Against this background, this chapter deals on a theoretical level with the significance of the interactions between biological and social factors for the epidemic or pandemic spread of infectious diseases. This biosocial synthesis also serves to describe vulnerability more precisely both as a biological property and as a social property (e.g., belonging to a social group that is frequently exposed to pathogens). The purpose of this chapter is to roughly organise the field of research and to bring together previously known principles or hypotheses in one contribution. I use a classic model of infectious disease epidemiology, namely, the epidemiological triangle (also called the triad), as a theoretical framework. It describes the causation of infectious disease as the product of an interaction between pathogen, host and environment. However, it is precisely the aspect of the interaction of the three factors that has, thus far, only been theoretically differentiated to a limited extent. This also applies to the interaction of the biological characteristics of the pathogen and host and the social environment in which both move. Accordingly, I draw on ecosocial theory, a concept from public health research that describes relationships between people's social environment and their health. To explore the possibility of linking the models, an exemplary analysis is presented in the second part of this chapter. If focuses on the question of what conclusions can be drawn about relevant social factors from the knowledge of the biological properties of pathogens alone. In other words, the biology of the pathogen is used to look at social factors (biology to social).

2 Conceptual Framework

Following the arguments outlined in the introduction, a theoretical model to describe biosocial relationships should be both disease-specific and integrate notions of biological and social processes. A classic model of the development of infectious diseases that fulfil these conditions is the epidemiological triangle. It goes back to work in plant biology and veterinary epidemiology and postulates that three variables and their interaction are crucial for infectious disease causation: characteristics of the pathogen, the host and the environment [9, 10].

First, I would like to outline the basic characteristics of the pathogen. Only some of the basic characteristics are explained here initially; a more differentiated consideration will follow later in the text. It should be noted that the central feature of the *aetiology* (development) of infectious diseases is the necessary presence of a specific biological *pathogen*. These are different forms of organisms, whereby the most important pathogens in human infectious diseases are viruses, bacteria, parasites, fungi and microbes [11]. The relevant pathogens are those that are able to *colonise* and *infect* a host—in this case, a human being [12]. While colonisation is superficial, infection is the penetration of the pathogen into the host tissue. For some pathogens, such as many viruses, infection also means that the pathogen manages to replicate initially in the host organism using its resources. When assessing how effectively and efficiently a pathogen spreads (transmission), a third question is crucial: which *mode of transmission* the pathogen uses to colonise and infect a host? This also includes a consideration of the circumstances under which and the intensity with which a host has contact with the pathogen, i.e., whether it is *exposed* to it. Once an infection has occurred, the further course of the infection is also pathogen specific. There can be symptom-free courses with complete healing and symptomatic diseases up to chronic or fatal courses. Whether a disease develops from an infection is determined by the pathogen's individual *pathogenicity* (the ability of the pathogen to trigger a disease). The severity of the disease, including the risk of a fatal course, is related to the *virulence* of the pathogen [13].

The second edge in the triangle is the host. First, the host determines through its behaviour (a) whether an exposure occurs at all and (b) if a superficially colonising pathogen leads to an infection, for example, by preventive behaviour, such as hand hygiene. A further host-related factor that is decisive for disease transmission is the so-called *susceptibility* of the host, which denotes the probability that a host becomes infected if exposed to a pathogen [14]. This is determined by the immune system and the general constitution or resilience of the organism. Both have a genetic component but are also flexible. The immune system, for example, is capable of adapting to certain pathogens and developing immunity through previous infections or external immunisation (i.e., vaccination) to improve its defense against subsequent exposures to the same pathogen [14]. Once infected, immunity and constitution remain important because they also influence the further course of the disease (in interaction with the specific pathogenicity and virulence of the pathogen). People with chronic preconditions, for example, may be particularly vulnerable to developing severe disease if an infection occurs.

The environment, as the third variable, influences both the pathogen and the host and determines how often and under what circumstances the two come into contact with each other. Strictly speaking, it is, thus, less to be understood as part of a triangle but rather as the framework within which host and pathogen exist and interact. The environment was originally understood primarily as the natural environment in the epidemiological triad [10]. Most pathogens need very specific environmental conditions to develop and spread. Important examples are temperatures and their seasonal stability, solar radiation, the presence of certain habitats or biotopes (e.g., stagnant water as a breeding ground for certain mosquitoes) and the presence of host animals in the case of pathogens whose natural reservoir is animals (see below). These conditions are both of current importance for existing pathogen populations and formative in the long term, as they govern the evolutionary adaptation and genetic development of these organisms.

The environment in the epidemiological triangle also includes the social environment [13]. A generally valid and theoretically sound taxonomy of the social factors that are relevant in this context does not yet exist. However, Van Seventer and Hochberg have presented a list of broader categories in their presentation of the principles of infectious diseases that are used here [13]. The categories "land use", "infrastructure" and "technology and industry" are directly related to the natural environment. Land use covers a wide spectrum of human interactions with nature, from agriculture and forestry to water resource management. Although Seventer and Hochberg do not explicitly mention it, it can also include the built environment, i.e., settlements, cities, industrial areas, etc. Land use is a term that is used to describe a wide range of human activities. The type and intensity of land use not only change the natural environment and thus the living conditions for pathogens but also determine the ecosystems in which people spend time and possibly come into contact with pathogens. Under infrastructure, the authors sum up particularly those technical infrastructures that are connected to pathogen habitats. These can be, for example, defective water pipes on a small scale or systems for water treatment and supply of entire regions on a large scale. Technology and industry also represent a broad category that includes many subcategories from medical technology (e.g., antibiotics, disinfection) to food processing and industrial animal husbandry. Another category is "travel and commerce", i.e., the movement and encounter of people and goods that are potential reservoirs for pathogens (e.g., import of animals). "Economy" is also listed as a factor, although Van Seventer and Hochberg define it quite narrowly and focus primarily on economic inequality and its consequences for access to hygiene and clean water/food. Closely related to this is the category "war and conflict". Refugee movements, unhygienic living conditions or a lack of access to clean food are typical consequences of armed conflicts and make it easier for pathogens to spread in the populations affected. "Policy" is also on the list but is somewhat narrowed to policy interventions aimed directly at controlling outbreaks. The final category listed is "social/behavioural" norms and culture. Sociocultural factors shape individuals' contact or hygiene behavior and, therefore, have a major impact on infection risks [1]. Examples of significant norms are conventions for distance or closeness in everyday encounters, hygiene norms (e.g., regarding hand washing), sexual norms or norms for illness behaviour in the case of a symptomatic illness.

This categorisation of social influences on infectious diseases is not complete. The health care system, for example, and its effectiveness in preventing and treating disease is missing. However, Seventer and Hochberg do not claim to be exhaustive. Their aim is rather to illustrate the importance of social structures and practices and identify important problem areas. Naming individual social factors of influence, however plausible their significance may be in individual cases, can only be a first step. These are not disjunctive categories but, instead, highly interdependent variables. Policy, for instance, influences almost all other categories but is itself also influenced by them (e.g., technical progress can make regulatory action necessary). Furthermore, it is necessary to define and specify idiosyncratic interactions with the other two angles of the triangle, i.e. pathogen and host, for the various social categories.

To grasp the complexity of the social environment and biosocial interactions within the epidemiological triangle better, it might be useful to make references to general theoretical ideas on the social determinants of human health. Nancy Krieger's ecosocial model [15, 16] is of particular interest in the case of infectious diseases. It builds on more general models of the social determinants of health and extends them by including ecological, epidemiological and medical-sociological ideas of disease causation. First, it postulates that influences on health can manifest themselves on different social levels, from the micro level of social actors to the meso level of social networks and social environments (e.g., households, neighbourhoods, work, consumption/markets, education) to the macro level of social structures and systems (e.g., economic, political, health and legal system). Although the levels are interdependent, they can each be conceptualised as specific environments that influence people's health-related behaviour in a characteristic way, as well as containing material and psychosocial risks and resources for health. Another important element of the model is the assumption that there is a modification of influences of the social environment by the political economy of a society. This refers to systems of an unequal distribution of resources (e.g., health, power, access to cultural capital, distribution of income and wealth), which are linked particularly to the characteristics of gender, race/ethnicity and social class. Structural distributional inequalities lead to health inequalities in the next step, as socially disadvantaged individuals are more often exposed to potentially harmful environments than better-off people. This structural element is highly significant, as almost all diseases, including infectious diseases, more often affect disadvantaged populations than more privileged populations (health inequalities) [17].

What distinguishes the ecosocial model from related models on the social determinants of health is the explicit consideration of the biosocial processes that link influences of the social environment and human health. According to Krieger, confrontation with the social environment leads to a so-called *embodiment*—understood as a translation of environmental influences into pathogenic processes. In this context, Krieger speaks of "pathways of embodiment […], that involve exposure, susceptibility, and resistance (as both social and biological phenomena)" [16, p. 225]. Embodiment can thus be understood as the interface between biology and the social environment and is therefore of paramount importance for understanding the genesis of diseases. If these processes are understood, it is possible to assess which social factors can be considered influences in the first place and which biological factors make people particularly susceptible to social influences. If embodiment is interpreted in this way, however, it is necessary to look at this interface in a disease-specific way, since, as is well known, every disease has its own pathology.

However, the ecosocial model has not yet been adapted for the specifics of infectious diseases—even though direct references to the epidemiological triad can be seen in the concept of "pathways to embodiment". In contrast to chronic degenerative diseases, where exogenous risk factors that arise in the social environment (e.g., noise pollution in a noisy workplace) have an effect on the person and influence their health, in the case of infectious diseases, the pathogen and its peculiarities must be considered an "actor", so to speak. The social environment is then no longer just something that affects a person via mediators (i.e., risk factors) but can also be, for example, the physical habitat for the pathogen. Exemplarily, a virus that is transmitted directly from person to person spreads particularly effectively in social contexts in which as many people as possible come together physically, be it in large gatherings or through the contacts of mobile populations distributed over a large area. If it succeeds in infecting a large number of people, it becomes part of the social environment itself and can, in turn, infect other, previously unaffected people. At the same time, the social environment shapes the resilience of the members of a community, for example, via the influence that living and working conditions have on the general state of health or the targeted build-up of immunity to pathogens through socially organised vaccination campaigns. Thus, a continuous and dynamic interaction between pathogens, hosts and the social contexts in which they are embedded can be assumed [1, 18]. This idea of interaction is also taken up by Merrill Singer and colleagues in their syndemic concept [19]. They focus on explaining the phenomenon that certain diseases tend to occur together (clustering). However, Singer and colleagues explicitly state that interactions between diseases occur because "contextual and social factors create the conditions in which two (or more) diseases or health conditions cluster" [19, p. 942].

In sum, linking the epidemiological triangle with ecosocial theory could provide a suitable theoretical framework to hypothesise specific roles of social factors in the spread of infectious diseases. As mentioned, the model needs to be adapted to the specificities of infectious diseases. This is not possible in its entirety here due to the complexity of the interactions indicated. Instead, the following section will take a first step by taking a closer look at the process of embodiment in infectious diseases and drawing consequences for the analysis of social environments.

3 Characteristics of Infectious Diseases: Pathways to Embodiment

There are a number of characteristics that distinguish infectious diseases from noncommunicable ones. Knowledge of these is relevant in that the biological characteristics help determine which social influences may be important at different stages of the disease process. In this section, some principles of the biology of infectious diseases are listed to identify possible interfaces with social conditions that make it easier or more difficult for the pathogen to spread. I follow the biological process from exposure to infection to disease and, step-by-step, address pathogen characteristics and their implications for interaction with hosts and social contexts. The explanations are general, make no claim to completeness and are not related to a specific pathogen. To make connections clear, however, COVID-19 is repeatedly used as an example.

4 Transmission: Modes of Transmission

The process begins with the colonisation and infection of a host by the pathogen. The pathway by which this occurs is highly interesting regarding the analysis of social influences, as it is usually directly associated with host behaviour. There are two basic mechanisms: direct transmission of the active pathogen from host to host and indirect transmission of the pathogen via the intermediate step of colonisation or infection of a vehicle [20]. Many pathogens are able to transmit via both routes simultaneously, but the efficiency of transmission may differ depending on the respective route [21].

The two routes can be differentiated further, which will be done here, first, for direct transmission. In this case, there are various modes of transmission, which are generally based on an infected person releasing contaminated material via a portal of exit (e.g., respiratory tract) and this being taken up by another person via a portal of entry [20]. Carriers are primarily body fluids, such as droplets or aerosols, which are released with the respiratory air, blood, breast milk, tear fluid, sweat, semen, vaginal secretions, urine or stool. Which carrier plays a role depends on the pathogen. According to current knowledge, SARS-CoV-2, for example, spreads primarily via secretions that are released with the breath [22, 23]. Droplets, i.e., larger particles of salivary secretion that contain active virus material, are of particular importance. They are produced in larger quantities primarily when speaking (or singing), coughing or sneezing. Due to their size, however, they sink to the ground relatively close to the infected person within a radius of only a few feet [23]. This is important for transmission because infection or initial colonisation by the virus occurs primarily in the upper respiratory tract of the new host. In this respect, a certain physical proximity of one to two metres to the infected subject is necessary to inhale droplets. However, smaller particles (aerosols), which can remain airborne for longer periods and over longer distances, also contain virus material and could initiate infections, especially in closed rooms [24]. In addition, other modes of transmission are also discussed, for example, contact infection through active viruses in stool or urine, which, however, only seem to play a minor role overall [23].

If the transmission properties of the pathogen are known, conclusions can be drawn about social situations that make exposure to a pathogen likely. In the case of SARS-CoV-2, for example, situations in which people come together without much distance and hold conversations or sing/play music together, as well as long stays in unventilated rooms, would be risky. These general considerations allow the identification of concrete social situations in which the risk of transmission is presumably increased. In the case of COVID-19, these would be, for example, visits to concerts with large audiences in confined spaces, working in open-plan offices, meetings in confined spaces and school lessons in crowded classrooms without ventilation. Thus, knowledge of the biological mechanisms involved in the transmission of a pathogen alone allows hypotheses to be made about social contexts with exposure potential. Many public health interventions to control infections are based on such hypotheses. The discussions about nonpharmaceutical interventions must be seen against this background, especially at the beginning of the pandemic when it was still relatively unclear which modes of transmission were important and there was, therefore, a great deal of uncertainty about which measures should be taken [25].

In indirect transmission, on the other hand, the pathogen spreads from one host to another via a so-called vehicle. There are many possible vehicles of transmission. Some pathogens are transmitted through another organism (so-called vector-borne transmission), while others use all possible forms of inanimate matter (e.g., water-borne, airborne, food-borne, object-borne) [21]. Examples are the vector-borne transmission of plasmodia via mosquitoes (Anopheles gambiae) in the case of malaria or the water-borne transmission of the bacterium Vibrio cholerae via contaminated water in the case of cholera [14]. It should be noted that such transmission routes are not always "true" indirect transmissions in the form of a host-vehicle-host chain. There are also pathogens (such as certain bacteria) that permanently colonise inanimate matter and, from there, colonise one host after another. [20] Another variant is that the original host is a vector per se. This applies, for example, to zoonoses, i.e., infectious diseases that can be transmitted from animals to humans. As in the case of direct transmission, knowledge of the vehicles or possible intermediate steps in the infection chain allows general conclusions to be drawn about potentially risky social situations and practices. In the case of vectors, i.e., animals as the source of infection, the characteristics and behaviour of the respective vector become a further relevant factor of transmission in addition to the pathogen properties. In the case of the malaria mosquito mentioned already, staying outdoors or in open spaces during the vector's active times and without mosquito protection are risky. Zoonoses also affect the relationship between animals and humans, such as the forms and intensity of animal husbandry in agriculture or the conditions under which pets are kept. In the case of transmission via the environment, on the other hand, all of the human everyday behaviour around nutrition and hygiene (e.g., personal hygiene, cleaning of surfaces) potentially plays a role.

5 Transmission: The Natural Environment

Transmission must be seen in the context of the natural environment and its diverse physical and biological phenomena. The pathogens, their vectors and the hosts move within this environment. Pathogens are generally dependent on certain environmental conditions for their spread or are only able to survive under certain conditions. This applies both to their modes of transmission and their general ability to reproduce, for example, in a preferred vehicle. Which ecosystems pathogens prefer is highly individual. There are those that occur exclusively in small ecological niches (often parasites, such as the guinea worm), while other pathogens can potentially exist in very many ecosystems [14]. There is a dependence on environmental characteristics in both direct and indirect transmission. However, for pathogens that primarily transmit indirectly, there is also dependence on the presence of their preferred vehicle [26]. Therefore, the relationships of this vehicle to the environment and its preferred habitats and biotopes must also be considered here. A certain dependence on climatic conditions and their seasonal changes is also typical. The activity, for example, of many pathogens fluctuates along seasonal variations in temperature, rainfall and solar radiation [27, 28]. There are several reasons for this, for example, the availability of vehicles changes with local weather conditions (e.g., standing puddles as breeding sites for mosquitoes) [29]. In addition, the survivability of the pathogen outside of a host can also be influenced by factors such as solar radiation or heat. However, people's behaviour also changes with seasonal weather, which is important for both direct and indirect transmission: they consume certain foods seasonally, stay outdoors for shorter or longer periods of time, or change their mobility and contact behaviour depending on the environmental conditions. Exemplarily, more encounters take place indoors when it is cold than during periods of high temperatures. On the one hand, this makes the direct transmission of pathogens transmitted via droplets or aerosols more likely; on the other hand, it reduces the probability of infection by a vector whose habitat is outdoors (e.g., ticks).

It is not possible here to begin to comprehensively describe the complex connection between pathogens and the natural environment. The few aspects listed above must therefore suffice to make clear that the embedding of the pathogen in its environment also has consequences for its interaction with the social environment. the questions in which concrete habitats pathogen and host meet and which moderating influence environmental conditions have on the probability of exposure and infection are essential points. These considerations must also be seen in the context of the massive anthropogenic changes that have affected the entire planet since industrialisation and its fossil economy. Not only global climate change and a general change in land use but also socioeconomic change, for example, related to mobility or income distribution, are expressions of this [26]. This also changes the living conditions for pathogens, and it is assumed that there will be more infectious diseases in the future [30]. There are many reasons for this, such as global warming, which makes new geographical zones accessible to certain pathogens (e.g., malaria, dengue fever), or the increase in zoonoses due to the intrusion of immune-naive humans into previously untouched biotopes or through intensified animal husbandry [26, 29]. SARS-CoV-2 is also suspected to be a zoonosis, and the natural environment of the pathogen is in wildlife populations [31]. Only the intrusion of humans into these unspoiled habitats and the capture and sale of creatures for food purposes would subsequently have promoted the spill over.

6 Transmission: Timing

The last aspect to be addressed in the context of transmission is the time period in the course of the stages of infection and disease during which an infected host (source) can infect another host [13, 20]. This is particularly important in the case of direct human-to-human transmission. Once the source is infected, the pathogen usually needs some time to replicate and colonise a new host. The time it takes for an infected person to become infectious and pass on the pathogen is the latent period. This is followed by the period during which the person remains infectious (infectious period). Another variable is the incubation period, i.e., the time between infection and the appearance of the first symptoms of a disease. This is then followed by the period of clinical disease. The ratio of the respective times varies between pathogens and has a strong influence on the effectiveness of transmission. The time interval between the latent and incubation period and the overlap between the infectious period and the duration of a symptomatic disease are of particular importance. Asymptomatic carriers (the latency period is shorter than the incubation period or the infectious period is longer than the clinical period) are especially very efficient in passing on the pathogen. The reason is an interaction with social practices. The appearance of symptoms is an external sign of disease, causing infected people to reduce social contact either independently or at the instigation of third parties (e.g., quarantine rules, inpatient hospitalisation). Asymptomatic carriers, on the other hand, have no reason to avoid social contact and thus cannot actively prevent the possible infection of others. Knowledge of the latency and incubation periods is therefore essential for dealing with outbreaks in society. Only then is it possible to develop appropriate strategies. If, for example, it is known that transmission occurs primarily via asymptomatic carriers, it is imperative to react at the societal level, for example, by carrying out screening tests or imposing contact restrictions for entire populations at risk. In the case of pathogens where the infectious period overlaps strongly with the clinical illness, on the other hand, it is important to educate people to avoid social contact when symptoms appear. In addition, in such a scenario, special attention must be paid to the protection of those who care for the sick.

7 Infectivity and Immunity

As indicated above, an additional distinction must be made between exposure, colonisation and infection. The exposure and subsequent colonisation do not necessarily mean that the pathogen also succeeds in infecting the person affected. This is determined by an interplay between the characteristics of both the pathogen and host. First, pathogens differ in terms of their *infectivity*, i.e., their general ability to infiltrate the organism of the host. Both the biological mechanism of infiltration and the dose of exposure required to enable the pathogen to infiltrate are decisive. SARS-CoV-2, for example, has high infectivity according to current knowledge. The virus, especially in its later variants (Delta, Omicron), infiltrates cells highly effectively by docking with its spike protein to cell receptors (i.e., angiotensin-converting enzyme 2 receptor: ACE2) and subsequently use the protein production of the cell to replicate and establish itself in the body of the new host [32]. However, the success of an infection also depends on the viral load to which the person is exposed [33]. This means that an infection becomes more likely if a high amount of virus is ingested in contact with infected people who excrete high amounts of virus (e.g., during particularly infectious stages of the disease or in the case of "superspreaders" who excrete an above-average viral load) [23].

However, the pathogen's ability to infect now comes up against the host's defense, which, in the context of infectious diseases, is referred to as *susceptibility*, i.e., the individual probability of becoming infected when exposed to the pathogen. [14] On a biological level, susceptibility is determined primarily by the immune system (although other factors also play a role, but this would go too far here). The human immune system is highly complex and can only be described here in a general way. What is important is that it has both a general, innate component and a variable, adaptive component. The innate immune system is not very specialised and operates against all pathogens it can recognise. Numerous processes are involved, such as the initiation of inflammation or the destruction of infected body cells by natural killer cells. The second component is the adaptive immune system, which is capable of learning and developing specific responses for individual pathogens. Immune memory is built up by previous infections with this pathogen or external immunisation (i.e., vaccination). If contact occurs again, the pathogen can be recognised, and an immune response tailored to it can be initiated, for example, by killer T cells [14]. Nevertheless, there is marked variance between hosts regarding the effectiveness of both the innate and adaptive immune responses, which means that some people are more susceptible than others. The variance may be the result of normal genetic diversity, but often the effectiveness of the immune system is associated with other characteristics. One key factor is age. The adaptive immune system, for example, must first be trained; thus, newborns or infants naturally have weaker defences against pathogens. The functionality of the immune system in older age then decreases again as a result of ageing processes, which also tends to make older people more susceptible. In addition, the immune systems of men and women differ slightly, which can lead to a different susceptibility of the biological sexes depending on the pathogen.

Such a difference has been observed with SARS-CoV-2. The T-cell response of women to the virus seems to be more robust than that of men; consequently, the infection rates of men are higher than those of women [34]. However, the immune system is not only determined by biological characteristics but also influenced by external social factors. One of these, chronic stress, has been shown to affect the immune system negatively [35, 36]. Since chronic stress is predominantly a product of social stressors, such as confrontation with aggressive behaviour, the experience of poverty and exclusion or poor working conditions, there are direct social influences on immunity here. Indirect immunisation through vaccination is also genuinely socially mediated, since vaccinations have to be developed by scientific institutions and made accessible by both the political and health systems.

Furthermore, the immune system is also influenced by the general constitution or resistance of the organism. Preexisting conditions such as metabolic disorders (e.g., diabetes) can weaken the immune system. The same applies to patients with diseases that are treated by specifically reducing the immune system with medication. Immunosuppressants, for example, are used in the treatment of various tumours, resulting in cancer patients becoming susceptible to infections. It should be noted that the general constitution is also determined to a great extent by social living and working conditions. Most widespread chronic diseases have a multicausal aetiology, and their development usually involves numerous social factors, such as socioculturally shaped lifestyles, stress or material disadvantage.

8 Pathogenicity and Virulence

If an infection occurs, immunity and constitution remain important, as they also influence the course of the infectious disease. Two characteristics of the pathogen are decisive here. Pathogenicity refers to the probability that an infection will also result in a symptomatic disease, while virulence refers to the severity or danger of a disease [20]. There is a high variance between different pathogens for both factors. Some of them rarely cause clinical disease, while others almost always lead to symptoms when infected. The dangerousness also differs. The most drastic indicator is the probability of a fatal course (case fatality). This is very low for pathogens such as normal cold viruses (e.g., rhinoviruses), while other pathogens, such as Ebola, have very high lethality.

Influences on the resistance of the hosts and the role of social factors have already been discussed in the previous section. However, at this stage of the disease, there are further interactions with the social environment. The pathogenicity of a pathogen is important in several respects for the social reactions to a disease. First, it determines how the social life of an affected society changes. If the pathogenicity is high and, simultaneously, the number of people infected is considerable, a large number of people will fall ill at the same time during an outbreak. On the one hand, this increases the need for medical care and, thus, the pressure on the health system. On the other hand, these infected people can no longer pursue their normal social roles for the duration of their illness. This can have consequences in many areas of society, especially if employees are absent in large numbers and economic life is disrupted as a result, or if care work is no longer performed and has to be compensated for in some other way. During the COVID-19 pandemic, for example, essential workers were temporarily affected by staff shortages. These were not only particularly health professionals but also employees in the transport sector and the food industry [37, 38]. Such outages can threaten the security of supply and health care and have further negative consequences for the health of the population. The lethality of a disease is also a pathogen characteristic that determines societal responses. It is obvious that the willingness to implement far-reaching measures to protect against infection will be significantly stronger in the case of a particularly lethal disease than an infection with only low lethality.

9 Conclusions

Biosocial interactions between the biological characteristics of pathogens or infectious diseases and social factors have been discussed in this chapter. However, the theoretical study of the biosocial basis of the spread of infectious diseases is not an end in itself. The COVID-19 pandemic has clearly shown that measures taken to prevent and control pandemics work, above all, if they interrupt chains of infection precisely and reduce susceptibility and vulnerability. The selection of such measures depends on the precise knowledge of both the biological properties of the pathogen and the social framework of embodiment. In preparation for future pandemics, it may therefore be useful to further develop the theoretical framework and test it empirically in a targeted manner.

The proposal was made here to fall back on a classic model of infectious disease causation but to combine it with a model of social determinants to be able to describe the interface between the biological properties of pathogens and hosts and the social environment in a conceptually better way. In my opinion, the theoretical analysis shows that Nancy Krieger's ecosocial theory is quite compatible with the assumptions of the epidemiological triangle. In this respect, it could be worthwhile in the future to develop an ecosocial model specifically for infectious diseases. This was only possible here as an example, and numerous possible systematics could just be hinted at. It would be desirable, for example, to further develop the taxonomy of relevant social contexts or of individual social factors on the occurrence of infections. To date, only general, incomplete category lists are available. In addition, the modes of action of the respective context factors and their different significance depending on certain pathogen characteristics still need to be understood better and systematically described.

Furthermore, it could be useful to continue the research on the biosocial basis of pathogen transmission and thus to work out the pathways to embodiment more clearly. From the knowledge of the biological properties of pathogens, it is possible, in principle, to deduce which social practices, situations and structures are associated with the probability of exposure, susceptibility and vulnerability to critical courses. If it is possible to determine generalisable principles of biosocial interactions as a function of the biological properties of pathogens, this could make it easier in the future to estimate which spreading dynamics can be expected when a specific pathogen encounters certain social conditions. It would be, for instance, conceivable to draw up a kind of checklist of individual pathogens or pathogen groups properties that could provide orientation in the assessment of a current outbreak. This could also provide lessons for the early identification of social groups that are particularly vulnerable in terms of risk of infection and/or severe courses. During the COVID-19 pandemic, for example, socially disadvantaged people had a higher than average burden of disease because their specific living and working conditions made protective behaviours difficult and exposure to the virus more likely (e.g., cramped housing, need to use public transport). Although this health inequality was uncovered by empirical studies relatively early in the pandemic, it was initially little noticed by the public in many countries; consequently, only sporadic policy measures were taken to protect these particularly vulnerable groups [8]. Therefore, systematic preparation for pandemics, which is also oriented towards the biosocial basis of the spread of infections, seems necessary to prevent such inequalities in the future.

Core Messages

- In the spread of infectious diseases, specific interactions between the biological properties of the pathogen and the host, as well as the social context in which this interaction takes place, play an important role.
- In order to describe biosocial interactions theoretically, classical concepts of epidemiology can be linked with public health theories. Plausible connections arise, for example, between the epidemiological triangle model and the ecosocial theory.
- Vulnerabilities can be both biological (e.g. an increased biological susceptibility of a person to become infected with a specific pathogen) and socially based (e.g. social disadvantage). As both aspects interact, a multidimensional conception of vulnerability seems to be necessary.

A continuous and dynamic interaction between pathogens, hosts and the social contexts in which they are embedded can be assumed.

Nico Dragano.

References

1. Arthur RF, Gurley ES, Salje H et al. (2017) Contact structure, mobility, environmental impact and behaviour: the importance of social forces to infectious disease dynamics and disease ecology. Philos Trans R Soc Lond B Biol Sci 372. https://doi.org/10.1098/rstb.2016.0454
2. Butler-Jones D, Wong T (2016) Infectious disease, social determinants and the need for intersectoral action. Can Commun Dis Rep 42:S118-S120. https://doi.org/10.14745/ccdr.v42is1a04
3. Kelly-Irving M, Delpierre C (2021) Framework for understanding health inequalities over the life course: the embodiment dynamic and biological mechanisms of exogenous and endogenous origin. J Epidemiol Community Health 75:1181–1186. https://doi.org/10.1136/jech-2021-216430
4. Kerrigan D, Donastorg Y, Barrington C et al (2020) Assessing and addressing social determinants of HIV among female sex workers in the Dominican Republic and Tanzania through community empowerment-based responses. Curr HIV/AIDS Rep 17:88–96. https://doi.org/10.1007/s11904-020-00485-3
5. Benita F, Rebollar-Ruelas L, Gaytán-Alfaro ED (2022) What have we learned about socioeconomic inequalities in the spread of COVID-19? A systematic review. Sustain Cities Soc 86:104158. https://doi.org/10.1016/j.scs.2022.104158
6. McGowan VJ, Bambra C (2022) COVID-19 mortality and deprivation: pandemic, syndemic, and endemic health inequalities. Lancet Public Heal 7:e966–e975. https://doi.org/10.1016/S2468-2667(22)00223-7
7. Bambra C (2022) Pandemic inequalities: emerging infectious diseases and health equity. Int J Equity Health 21:6. https://doi.org/10.1186/s12939-021-01611-2
8. Sirleaf EJ, Clark H (2021) Report of the independent panel for pandemic preparedness and response: making COVID-19 the last pandemic. The Lancet. https://doi.org/10.1016/S0140-6736(21)01095-3
9. Snieszko SF (1974) The effects of environmental stress on outbreaks of infectious diseases of fishes. J Fish Biology 6:197–208. https://doi.org/10.1111/j.1095-8649.1974.tb04537.x
10. Scholthof K-BG (2007) The disease triangle: pathogens, the environment and society. Nat Rev Microbiol 5:152–156. https://doi.org/10.1038/nrmicro1596
11. Alberts B, Johnson A, Lewis J et al (2002) Molecular biology of the cell, 4th edn. Garland Science Taylor & Francis Group, New York
12. Fox MW, Gower EW (2021) Infecious disease epidemiology. In: Lash TL, VanderWeele TJ, Haneuse S et al. (eds) Modern epidemiology, 4th edn. Wolters Kluwer, Philadelphia, Baltimore, New York, Buenos Aires, Hong Kong, Sydney, Tokyo, pp 805–843
13. van Seventer JM, Hochberg NS (2017) Principles of infectious diseases: transmission, diagnosis, prevention, and control. Int Encyclo Public Health, pp 22–39. https://doi.org/10.1016/B978-0-12-803678-5.00516-6
14. Abubakar I, Stagg HR, Cohen T et al (eds) (2016) Infectious disease epidemiology, 1st edn. Oxford specialist handbooks in infectious diseases. Oxford University Press, Oxford
15. Krieger N (2001) Theories for social epidemiology in the 21st century: an ecosocial perspective. Int J Epidemiol 30:668–677. https://doi.org/10.1093/ije/30.4.668
16. Krieger N (2008) Proximal, distal, and the politics of causation: what's level got to do with it? Am J Public Health 98:221–230. https://doi.org/10.2105/AJPH.2007.111278
17. Bartley M (2017) Health inequality: an introduction to concepts, theories and methods, 2nd edn. Polity, Cambridge, UK, Malden, MA, USA
18. Bonds MH, Keenan DC, Rohani P et al (2010) Poverty trap formed by the ecology of infectious diseases. Proc Biol Sci 277:1185–1192. https://doi.org/10.1098/rspb.2009.1778
19. Singer M, Bulled N, Ostrach B et al (2017) Syndemics and the biosocial conception of health. The Lancet 389:941–950. https://doi.org/10.1016/S0140-6736(17)30003-X
20. Krämer A, Akmatov M, Kretzschmar M (2010) Principles of infectious disease epidemiology. In: Krämer A, Kretzschmar M, Krickeberg K (eds) Modern infectious disease epidemiology: concepts, methods, mathematical models, and public health. Springer, New York, pp 85–99

21. Giesecke J (2017) Modern infectious disease epidemiology, 3rd edn. CRC Press, Boca Raton
22. Handiso TB, Jifar MS, Nuriye Hagisso S (2022) Coronavirus's (SARS-CoV-2) airborne transmission. SAGE Open Med 10. https://doi.org/10.1177/20503121221094185
23. Meyerowitz EA, Richterman A, Gandhi RT et al (2021) Transmission of SARS-CoV-2: a review of viral, host, and environmental factors. Ann Intern Med 174:69–79. https://doi.org/10.7326/M20-5008
24. Baselga M, Güemes A, Alba JJ et al. (2022) SARS-CoV-2 droplet and airborne transmission heterogeneity. J Clin Med 11. https://doi.org/10.3390/jcm11092607
25. Leclerc QJ, Fuller NM, Knight LE et al (2020) What settings have been linked to SARS-CoV-2 transmission clusters? Welcome Open Res 5:83 https://doi.org/10.12688/wellcomeopenres.15889.2
26. Hassell JM, Newbold T, Dobson AP et al (2021) Towards an ecosystem model of infectious disease. Nat Ecol Evol 5:907–918. https://doi.org/10.1038/s41559-021-01454-8
27. Grassly NC, Fraser C (2006) Seasonal infectious disease epidemiology. Proc Biol Sci 273:2541–2550. https://doi.org/10.1098/rspb.2006.3604
28. Rohr JR, Cohen JM (2020) Understanding how temperature shifts could impact infectious disease. PLoS Biol 18:e3000938. https://doi.org/10.1371/journal.pbio.3000938
29. Rossati A (2017) Global warming and its health impact. Int J Occup Environ Med 8:7–20 https://doi.org/10.15171/ijoem.2017.963
30. Watts N, Amann M, Ayeb-Karlsson S et al (2018) The Lancet Countdown on health and climate change: from 25 years of inaction to a global transformation for public health. The Lancet 391:581–630. https://doi.org/10.1016/S0140-6736(17)32464-9
31. Holmes EC, Goldstein SA, Rasmussen AL et al (2021) The origins of SARS-CoV-2: a critical review. Cell 184:4848–4856. https://doi.org/10.1016/j.cell.2021.08.017
32. Da Silva SJR, Kohl A, Pena L et al (2023) Recent insights into SARS-CoV-2 omicron variant. Rev Med Virol 33:e2373. https://doi.org/10.1002/rmv.2373
33. Barreto ML, Teixeira MG, Carmo EH (2006) Infectious diseases epidemiology. J Epidemiol Community Health 60:192–195. https://doi.org/10.1136/jech.2003.011593
34. Takahashi T, Ellingson MK, Wong P et al (2020) Sex differences in immune responses that underlie COVID-19 disease outcomes. Nature 588:315–320. https://doi.org/10.1038/s41586-020-2700-3
35. Austin MK, Chen E, Ross KM et al (2018) Early-life socioeconomic disadvantage, not current, predicts accelerated epigenetic aging of monocytes. Psychoneuroendocrinology 97:131–134
36. Marques AH, Bjørke-Monsen A-L, Teixeira AL et al (2015) Maternal stress, nutrition and physical activity: impact on immune function, CNS development and psychopathology. Brain Res 1617:28–46. https://doi.org/10.1016/j.brainres.2014.10.051
37. Lyttelton T, Zang E (2022) Occupations and sickness-related absences during the COVID-19 pandemic. J Health Soc Behav 63:19–36. https://doi.org/10.1177/00221465211053615
38. Reuter M, Rigó M, Formazin M et al (2022) Occupation and SARS-CoV-2 infection risk among 108 960 workers during the first pandemic wave in Germany. Scand J Work Environ Health 48:446–456. https://doi.org/10.5271/sjweh.4037

Nico Dragano is Professor of Medical Sociology at the University Hospital Düsseldorf, Germany. His research focuses on occupational health and the connection between people's social situation and their health opportunities. This research aims to identify social, cultural and economic influences on health. Examples are the study of the medical effects of poverty or the research of health-endangering stress in the working world. Building on the results of basic research, his working group also investigates how the health of people in difficult social and economic situations can be better protected through measures of prevention and health promotion. He has many years of experience in (social) epidemiological research on the interrelation of social and biomedical factors using large data sets from population-based studies. He is currently involved in several studies on socio-economic risk factors for COVID-19 infections and major diseases. He is a member of the coordinating committee of the German Health Network COVID-19.

Current publications can be found here: https://www.uniklinik-duesseldorf.de/patienten-besucher/klinikeninstitutezentren/institut-fuer-medizinische-soziologie/das-institut/publikationen or here: https://twitter.com/MedSozDUS

Part II
Who Is Vulnerable?

Chapter 6
Vulnerability and Gender After COVID-19

Sandra Laugier

First of all, focusing on the concern with vulnerability and insecurity can itself be valuable in bringing an important perspective to the attention of the world.
Amartya Sen, Human Security Now, 2003

Abstract The COVID-19 pandemic was a global tragedy, but it also presented a strange pedagogical moment. Women have long been the very name of what has been neglected and despised by public policies, and the consequences of this are also the lack of attention (the lack of care) paid by governments over the last decade to all the sectors in charge of the care and protection of citizens (health, education, poverty, old age, disability). The recent crisis related to the COVID-19 highlighted the vulnerability of human societies and their dependence on care and has shown that our usual conceptual tools are insufficient in the face of the loss of the form of life woven by daily care.

Keywords Vulnerability · Care · Ethics · COVID

1 Introduction

We sometimes imagine that we are out of the COVID-19 pandemic, and we evoke a return to normal life with relief. This fervour for normalcy is itself a pathology linked to the pandemic and reveals something about the nature of the period we think we are emerging from: a period of collective suspension, of global anxiety, of a fear of others; and, also, a time of denial, even blindness, regarding our vulnerability. Pandemic: we are affected as a *whole*; it is the very form of human life that is disrupted. This effect is extraordinary when we consider that the number of dead,

---

S. Laugier (✉)
Institut Universitaire de France, Institut des sciences juridique et philosophique de la Sorbonne UMR 8103, CNRS Paris 1, Université Paris 1 Panthéon Sorbonne, Paris, France
e-mail: sandra.laugier@univ-paris1.fr

S. Achella and C. Marazia (eds.), *Vulnerabilities*, Integrated Science 18,
https://doi.org/10.1007/978-3-031-39378-5_6

while still terrifying in numerical terms, is nonetheless insignificant when compared to the number of survivors. All those who inhabit and wander the earth are now those who survived, yet there is none of the heroic language that was commonly attached to the term *survivor* prior to the pandemic.

But what remains of the dead from the pandemic? The "Leave This Seat Free" signs that appeared in metros and subways around the world, signs that have now disappeared, seemed like a reminder, a conjuring of the place of the dead. How did we *perceive* this sickness? Vulnerability is not anymore merely a general and empty word: bare life is *life permanently exposed to death*, and the pandemic accentuates the persistent sovereign power to take life in a modern era where humanity already lives most of their lives at a distance. The pandemic appeared as a threat to human life, both in the vital sense (a risk to biological life) and in the social sense (a risk to societal life: the standard way societies are organized was disrupted by the suspension of activities, the lack of public transport, the closure of schools, etc.). The reality of the health crisis thus revealed radical vulnerabilities. The vulnerability of people, of institutions, and of normal, ordinary life. It is precisely the accepted normalcy that was threatened and rendered strange (*uncanny*, in the sense of disturbing). Ordinary life in the sense of everyday life was suddenly made visible, revealed through its absence, in the disappearance of what comprised it: anonymous interactions in the street, on buses, and in public places, and shared experiences (coffees, conversations, weddings, funerals). The entirety of the pan of pandemic is indeed this global vulnerability, the vulnerability of life, of the human species, and of the world. Vulnerability of the form of life itself, which was revealed by the universality of the threat but also with the sudden visibility of all those who take care of the living and the dead.

The COVID pandemic was a global tragedy, but it also presented a strange pedagogical moment. The importance of care and the people who take care of us appears to everyone, and ignorance on the part of an entire society of what *makes it live,* whether it be in daily life or in the urgency of the risk of death, became finally obvious. If such a moral education is possible, it is because the disaster has revealed radical vulnerabilities. The vulnerability of institutions, the vulnerability of the species; the vulnerability of fragile populations who are precisely on the front line but also the vulnerability of every individual brought back to their home and back to their own resources, without the myriad of people and services that accompany him—back to housework, tidying up, even schooling … to services usually entrusted to others. The *grammar of care* has thus subtly imposed itself on everyone because care is never so visible as in those situations where the normal form of life is shaken [see 1, 2].

The word *care* has been at the centre of the global conversation. Care-work has been revealed as what keeps everyone going. And what is least acknowledged. What matters most to ordinary, but also professional lives, what makes it possible?—the work of caregivers … but also cleaners, garbage collectors, cashiers, delivery people, truck drivers; and in fact everything that matters *least* in the scale of values.

In the exposure to disaster, the truth of our dependencies emerges. *We* are all vulnerable, dependent on others. It is indeed these two meanings of life, biological and social, that have suddenly imposed themselves on us: the life that is given to us (mainly by women) and that we can lose; the everyday life, made possible or helped (mainly by women). Women take care of our forms of life—"form of life" understood, to quote Stanley Cavell and Veena Das, both in a horizontal sense (our social life) and in a vertical sense (biological life). The continuum of care activities, so complex to explain in theory, has finally become clear—the care that makes us live extends from the hospital to the supermarket.

In the crisis, women are curiously omnipresent … and absent. Present on all fronts, because they are constantly shown to us in the media: at their sewing machines, making makeshift masks; at the broom, cleaning up in hospitals and stores that are still open; at the bedside of patients, whose well-being they ensure, whose lives they save; at the cash registers of the businesses that allow us to continue a normal life. A wave of collective bad conscience is emerging; customers greet and thank the cashiers as they pay for their purchases—cashiers to whom a few weeks ago they would not have given a glance because they were too busy speaking on their phone to someone not present but clearly much more important.

This is an awareness of care, of the role of women and other help in our daily lives. It is the work of care that at the moment ensures the continuity of life. Society must be defended, certainly. However, those who defend it are the invisible ones who, until recently, were taken for granted as the underwater face of society, the taken for granted that make our lives possible. Reduced (in whole or in part) to our domestic lives, we realize that we are in constant need of care … because suddenly, we are, each in our own way, men and women, at last, doing some of the work, the cleaning, the tidying up, the raising and schooling of children … work so often normally entrusted to others. In addition, in public life, we heroize the work of care, first in the form of the work done in the hospital; then in other, more modest forms.

Care is changed by COVID, but the concept also comes to prove its political relevance. The very grammar of care has been imposed on all of us: we are all dependent on others, whether for vital needs, for life and death, or for more ordinary needs. It is indeed the two meanings of life, biological and social, that suddenly impose themselves on us: the life that is given to us (mainly by women) and that we can lose; daily life, made possible or helped (mainly by women). Awareness of vulnerability is also what makes this new sensitivity possible. We are all vulnerable, even if not all in the same way or to the same degree, and this extends to our health risk.

Care is at once a practical response to specific needs and a sensitivity to the ordinary details of human life that *matter*. Hence, care is a concrete matter that ensures maintenance (for example, as conversation and conservation) and continuity of the human world and form of life. This is nothing less than a paradigm shift in ethics, with a reorientation toward vulnerability and a shift from the just to the important. Measuring the importance of care for human life requires first acknowledging the truth: that human life forms are fundamentally vulnerable, subject to failure. To pay attention to ordinary life is to become aware of its vulnerability—it is constantly threatening to dissolve or else to reveal itself to have been unreal all along, a mere fantasy. Human vulnerability is the original condition of the need for care—what needs to be taken care of and cared about. We may add here a connection between security/safety and vulnerability.

The perspective of care, by calling our attention to our general situation of dependence, is thus indissociably political and ethical; it develops an analysis of social relations organized around dependence and vulnerability—blind spots of the ethics of justice. In response to the original position described by Rawls, the perspective of *care* would tend to set this original condition of vulnerability as the anchor point of moral and political thought. Not a position on which to build an ideal theory or set principles, but the mere fact of vulnerability that appears in the difficulty of reality. This is something that is obvious in the contexts Veena Das's *Life and Words* accounts for, when violence destroys the everyday and the sense of life as defining the human.

Autonomy, so much vaunted by philosophers—and by feminists as well, and by politicians—turns out to be an optical illusion: the autonomy of some is made possible by the work of others.

Attention to the everyday, to what Veena Das calls *the everyday life of the human*, is the first step in caring: care is attention, and the ethics of care calls our attention to phenomena commonly unseen but that stand right before our eyes. Here the definition of care by Joan Tronto and Berenice Fisher has to be taken very seriously:

> In the most general sense, care is a species of activity that includes everything that we do to maintain, continue, and repair our world so that we can live in it as well as possible. That world includes our bodies, our selves, our environment, all of which we seek to interweave in a complex, life sustaining web. [3, p. 40]

The perspective of care by calling our attention to our general situation of dependence and to the danger of denying these connections is thus indissociably political and ethical; it develops an analysis of social relations organized around dependence and vulnerability—blind spots of the ethics of justice. Care is a practice, not a moral feeling or disposition: you see the world differently.

Care is everywhere, and it is so pervasive a part of human life that it is never seen for what it is: activities by which we act to organize our world so that we can live in it as well as possible. When we get down to the ways that we actually live our lives, care activities are central and pervasive. How different the world looks when we begin to take these activities seriously. The world will look different if we place care, and its related values and concerns, closer to the centre of human life [4, p. 14].

2 Lessons of Care

Although men were the most numerous among the sick, women are vulnerable in a larger sense, they are massively impacted by the financial consequences of the crisis and they are also the most exposed. In addition to the fact that they are mostly part-time workers and have to take material and mental responsibility (the mental load) for domestic tasks, they constitute the vast majority of single-parent family carers. Not to mention the indifference of politicy-makers towards the elderly who die by the thousands in institutions—because institutionalized old age concerns women above all. And the fact that it turns out that long COVID affects them more than others.

Hospitals have a large majority of women on staff, especially at the lowest levels of the hierarchy, who are actually on the front line against coronavirus. The proportion of women is still rising among employees in nursing homes, home care workers and day-care centres. Women are in the majority at checkouts in shops, pharmacies, supermarkets.

We are therefore in a position of huge ambiguity in relation to care: women's work is still underestimated and underpaid at the very moment when its importance emerges in the eyes of all. Care has long been the very name of what has been neglected and despised by public policies, and it is indeed the lack of attention (the lack of care) paid by governments over the last decade to all the sectors in charge of the care and protection of citizens (health, education, poverty, old age, disability) that have made the fight against COVID so difficult.

The first lesson of COVID is a sudden awareness of a reversal of values that has been accepted for decades and denounced from the outset by the ethics of care: the most truly useful professions are the least well paid and the least well regarded. What matters most for our ordinary lives—carers, cleaners, garbage collectors, cashiers, delivery men, truck drivers—is in fact what counts the least in a scale of values that we have collectively validated. It is not only a matter of the multiple structural injustices that the epidemic has highlighted between those who are in the comfort of second homes and those who are at work. It has to do with the lack of knowledge—the denial—by an entire society of what keeps it alive.

It is carework that ensures the continuity of social life. Joan Tronto has been praised for the political version of care that she has proposed to emphasize the activity of care and for not defining care by affects, avoiding its devalorizing association to the realm of feeling. However, what matters is the definition that she proposes: "We suggest that caring […] preserve and repair our world so that we can live in it as well as possible" [3, p. 40].

The ethics of care, by suggesting a new attention to the unexplored or neglected details of life, confronts us with our own inabilities and inattentions. In becoming political, what is at stake in ethics of care is epistemological: they seek to bring to light the connection between our lack of attention to neglected realities and the lack of theorization (or, more directly, the rejection of the theorization) of these social realities, rendered invisible. Tronto has suggested that the dyadic image of *care* (such as maternal face-to-face) to which Carol Gilligan remains attached is too narrow to allow the ensemble of social activities having to do with attentive care for others to be thought. She considers that the philosophical valorization of care must base itself not so much on a particularistic ethics but rather on an enlargement of the concept of action and a move towards a neutral anthropology. Gilligan's position was indissociable from a gendered anthropology: for her, the relationship to the self and to others as expressed in moral judgment took opposing directions for men and for women. However, according to Tronto, this position would logically lead to a sort of anthropological separatism. She proposes instead an anthropology of *needs*, in order to find the social dignity of care: not only do certain of our needs (and among the most important ones) call directly for care, but care defines the (political) space in which listening to needs becomes possible, as a veritable attention to others.

We see that it is in passing from ethics to politics that ethics of care can be given their critical power. By calling for a society in which caregivers would have their voice, their relevance, and in which the tasks of care would not be structurally invisible or inconspicuous, they bring to light the difficulty of thinking these social realities. As Tronto puts it, the valorization of care passes through its politicization. The ethical affirmation of the importance and dignity of care cannot go without a political reflection on the allocation of resources and the social distribution of tasks this allocation defines:

> As a type of activity, care requires a moral disposition and a type of moral conduct. We can express some of these qualities in the form of a universalist moral principle, such as: one should care for those around or in one's society. Nevertheless, in order for these qualities to become a part of moral conduct, people must engage in both private and public practices that teach them, and reinforce their senses of, these moral concerns. In order to be created and sustained, then, an ethic of care relies upon a political commitment to value care and to reshape institutions to reflect that changed value. [5, pp. 177–178]6

Truly carrying out the ethics of care would imply both including practices linked to care in the agenda of democratic reflection and empowering those concerned—care givers and receivers. The recognition of the theoretical pertinence of ethics of care and the valorization of affects—the importance of which we have seen in correcting a narrow vision of justice—necessarily pass through a practical revalorization of activities linked to care and a joint modification of intellectual and political programs.

Caroline Criado Perez in *Invisible Women: Exposing Data Bias in a World Designed for Men* explains that 29 million articles were published about Zika and Ebola, but less than 1% of the publications concerned the gendered impact of the epidemic. Will we do better with COVID? While the current crisis highlights the importance of women's work in times of disaster [see the program:], it should also raise awareness not only of the essential role women around the world play in the production of the environment we live in but also of the risks to all of us from the invisibility of their contribution and the collective disregard for all the tasks of daily care and maintenance.

More than a change, it is an awareness of a reversal of values that has been accepted for decades and denounced from the outset by care analysts: the most genuinely useful jobs are the least well paid and the least highly regarded. What matters most for our ordinary life, what makes it possible—carers, cleaners, garbage collectors, cashiers, delivery men, truck drivers—called essential workers, where the word means actually they don't matter—in fact what counts the least in the collectively validated scale of values. It is not just a matter of the multiple structural injustices that the epidemic has highlighted between those who are in the comfort of second homes and those who are at work or crowded into transportation. It is a matter of the lack of awareness and denial by an entire society of what sustains it, whether in the flow of daily life or where there is an imminent risk of death.

Care is a critical concept. In fact, the empowerment of certain women through work and at the same time through the development of childcare systems, etc.—has also been achieved on the basis of a male model, in the sense that this autonomy has been achieved not by transferring tasks to men, or by a better distribution, but by putting other women at the service of women. So we don't want to ironize about these women who have become employers (and it is usually up to them to bear the moral and administrative burden of home-based employment, the mental burden) Rather, as is often the case in the care matters to show what is right under our noses: that the care tasks traditionally devolved to women still exist even if some women are (usually, in normal times) exempted from them. The crisis reveals to us that these tasks are just invisible; that they are usually taken up by immigrant and devalued populations, which again perpetuates the devaluation of care work and the moral categorizations that go with it.

Through the revelation it provides of the vulnerability of people, of all people, the perspective of care includes an ethical and political ambition, which is not only an active benevolence towards those who are close to us but constitutes *an education* in the perception and in the valorization of human activities. This is a reminder of the importance of rethinking care and outsourcing or service together. For the social—and today global—division of care work has until now risked giving the illusion that one can distinguish between emotional care—attentive to the emotional needs of particular people—and service care that can be delegated and purchased. The first would then be the prerogative of privileged white women, while the second remains restricted to everything that the former does not take care of, in short, the dirty work that is done by others. If the question of care is now bursting into the public sphere, it is also because the massive entry of women into the labour market has put traditional ways of providing care in crisis, but it is also because the confinement and the current restrictions put each woman back in front of this dirty work. It no longer works to outsource it for it to disappear. Whether it is provided within the domestic sphere or by public institutions or the market, care is produced at low cost by women whose social positions remain mostly precarious. Nurses, home helps, care assistants, social workers … not to mention all these other care professions that are devaluing at the speed of their feminization: teachers, doctors, judges, etc.

The care crisis is therefore both that of traditional caregivers, who are taking on an increasingly heavy burden due to longer lifespans, and that of the increasingly difficult conditions in which care activities are carried out, difficulties that have arisen as a result of the social policies that govern them in hospitals, institutions and private homes. This is the limit of the rhetoric on the valorisation and even empowerment of care workers. Today, at the moment of the crisis, we can look back at decades of neglect and malfeasance, which have ignored and wasted the natural and human resources that support humanity. Research on the role of women in agricultural work, in the management of resources or biodiversity, and in the preservation of daily lives are all avenues for clarifying questions of environmental justice and for perceiving the limits of a development concept designed to preserve Northern lifestyles based on the over-exploitation of environments, animals, and the populations of the South. The challenge is now to fight the invisibilization of care work carried out by women, which goes hand-in-hand with the invisibilization of the exploitation of resources.

Arlie Hochschild has thus proposed the concept of global care chains [7] to refer to these networks by which women delegate the care of vulnerable persons to one another. The corollary to these chains of care has been the emergence of a large, worldwide market of care. From North to South, care chains are defined by care work, its invisibility, and its devaluation. Care work, previously done silently and for free by women inside the home, is now outsourced: someone outside the family is called in and (barely) paid. The large market for the care of others may be able to solve certain problems in the countries of the North, but it creates new ones in the countries of the South.

Domestic work is increasingly performed by immigrant and devalued populations, which again perpetuates the moral devaluation of care work and the moral categorizations that go with it. What the sociologist Lewis Coser, in a controversial expression, called in the 1970s the obsolescence of domesticity [8] (the disappearance of domestics in American society) and attributed to the rise of democratic values, turns out to be the social invisibilization of domestic work, which is either paid and taken on by workers in a situation of social and political vulnerability or done for free by women at home.

The ethics of care is a method to reveal the invibilization of care work when it is done for the benefit of women (hence the feminist resistance to the concept of care). This inequality between women is at the core of the care drain from poor to rich countries. Drain because in many countries of the South, there has been a shortage of caregivers, who for the most part have gone to work in the countries of the North, where their skills are highly appreciated. However, this leaves their countries of origin facing a chronic shortage of personnel in the health sector. In the Philippines, for example, nearly 10% of the population works overseas and sends money home, and nursing is one of the most popular jobs. Each year, approximately 13,000 nurses leave to work abroad. Migrant nurses have played a very important role in helping countries such as Spain, Britain, and Italy fight the virus. We cannot fail after the pandemic to note the chronic lack of caregivers in the north in the face of growing needs, a lack made manifest by the characteristics of immigration today: women from Asia, Eastern Europe, and Africa who have come to serve as nannies, all-around maids, and sex workers [9], leaving behind other dependent people, including their own children, with no one to hand them over to. This immigration, which reduces the deficit of care in rich countries, amounts to a loss of care for poor countries.

In a famous study, the sociologist Rhacel Parrenas has shown that in the Philippines, the mass emigration of women has resulted in a depletion of care resources that is capable of disrupting society [10]. The observation of this persistent asymmetry in relations between North and South reinforces our earlier remarks about the invisibility of what sustains societies. Having plundered the natural resources of the South, exploited its physical strength first through slavery and then through migrations linked to industrialization, the North now endeavours to extract emotional and affective wealth from it. Privileged women, rather than fighting for a more egalitarian distribution of domestic tasks, delegate them to other women inside the home. And the fact that the women responsible for care come from dominated ethnic groups reinforces the power relations between the races and contributes to keeping domestic work invisible.

3 Global Ethics Lessons

Moreover, if we believe the IPCC, and if we look at socio-environmental inequalities and vulnerabilities, women are the ones who will pay a great deal in terms of adaptation to climate change. On 8 August 2019, the IPCC has published a Special Report on Climate Change, Desertification, Land Degradation, Sustainable Land Management, Food Security and Greenhouse Gas (GHG) Flows in Terrestrial Ecosystems [see 11]. Even if there are very great uncertainties in terms of adaptation, largely dependent on political choices, Asia and Africa are projected to have the greatest number of people likely to be dispossessed by desertification and environmental change; women are the ones on whom this everyday disaster will weigh the heaviest.

With regard to adaptation to climate change, as the same IPCC report notes, increased droughts and water shortages will mainly affect women, who are the main collectors, users and managers of water in poor countries. Water scarcity may increase their workload and reduce their ability to devote their time to other tasks, such as education. The increase in climate-related epidemics, with COVID-19 being only one of many that will inevitably follow, will mainly impact women, who, as we can see today, spend much of their time caring for the sick and raising children. Finally, the erosion of biodiversity has an impact on women's work, which depends on crop diversity and the proximity of food resources to adapt to climate variability. Women farmers are responsible for half of the world's food production and produce between 60 and 80% of the food in most developing countries [12]. Similarly, women are essential in supporting households and communities and in implementing mechanisms for adaptation and resilience, as the drafters of the report on climate change and gender equality write [13].

It is not coincidence that an essential and seminal work on this subject is the work of a woman, namely *Silent Spring* by Rachel Carson. As early as 1962, Carson highlighted the deleterious effects of pesticides on the environment, natural life and bird noise—that is, its aesthetic and sensitive dimension—calling for immediate political responses. It was as a result of such work that DDT was banned in 1972 in the United States. The 1970s saw the emergence of important ecofeminist movements and works in different countries that highlighted the importance of the environment. In this sense, the environment has been an important cause and a triggering process for many feminist struggles. It is ecofeminists in the South who have revived environmental thinking, showing in a radical way how, in countries that suffer from the legacy of colonial domination that has powered their economic potential but degraded their environment, the environmental consequences of development have affected women more heavily.

In India, the Chipko movement in 1973 against deforestation and Vandana Shiva's work on food and agricultural work is widely acclaimed [14]. One of the conclusions that may be drawn from these different works is the need to read environmental justice movements in terms of gender vulnerability, especially in light of future disasters. A better understanding of the changing relationships between women and the environment and an analysis of the ways in which women contribute to relational approaches to environmental management is essential for the future. Neglecting gender vulnerability and the unequal dimension of access and decision-making rights would doom environmental conservation to failure.[1] Indeed, according to the OECD (Social Institutions and Gender Index, SIGI), only 37% of the 160 countries studied give women and men equal access to land ownership and use. It is therefore important to develop a reflection on all future risks: the inequalities before the crisis (epidemic or other), during the crisis, the impact of these inequalities on the management of the crisis, and the consequences of the crisis on these inequalities. Post-disaster management must inevitably include issues related to the existence of patriarchal systems.

The capability approach developed by Amartya Sen and Martha Nussbaum emphasizes the interest in integrating women in a reflection that links capacities for action and possibilities of access, for example, to land ownership, resources, and education. The environmental disasters predicted and of which the current crisis is perhaps only one version (ocean acidification, desertification, sea level rise, coastal erosion, extreme events, etc.) are amply described in the IPCC reports. Women are still the ones on whom this everyday disaster will weigh most heavily. It is in this sense that international organizations are constantly advocating policies to empower women, given their role in the resilience of local environments and communities.

Numerous reports and studies show that women are often poorer and therefore more vulnerable in times of crisis. Hurricane Katrina, which ravaged New Orleans in 2005, affected African-American women and their children first and foremost. More than 70% of those who perished in the 2004 Asian tsunami were women. In Sri Lanka, it was easier for men to survive the 2004 tsunami because men had the advantage of being able to swim and climb trees, skills that are only taught to boys. In 1991, the cyclone in Bangladesh killed 140,000 people. For the 20–44 age group, the mortality rate for women was 71 per 1000 compared to 15 per 1000 for men. With regard to adaptation to climate change, increased droughts and water shortages will mainly affect women, who are the main collectors, users and managers of water in poor countries. Water scarcity will increase their workload and reduce their ability to devote their time to other activities, such as education.

So thinking beyond the COVID crisis means also thinking beyond the European, and the global North ... The revelation of gender inequalities and care work is only one part of a long list of global gender inequalities that are exacerbated in times of disaster. Post-disaster management will necessarily be gendered, and women must have a voice in it. The current crisis is rich in lessons about how to take into account future risks and the consequences of women's invisibility. Integrating the voices of women—of minorities, of all those who keep society alive, and who have been sent to care for others, sometimes risking their own lives—in the definition of *what counts* is indeed a matter of democracy: it is a matter of broadening the public and integrating ordinary life into the substance of political concern; it is a matter of recognizing the competence of subaltern people, which benefits the privileged who mobilize them,

[1] The issue of equity is consubstantially associated with that of sustainable development, as [15] show.

more than ever today, at their service. It is as if we take for granted, *together*, the services provided by women and by nature, without ever giving them a voice nor taking them into consideration in political decisions. Here, we touch on the crucial issue of social and political invisibility: it is a denial of services rendered.

The disasters of the last decades, and the most recent one, have highlighted the vulnerability of human societies and their dependence on care and have shown that our usual conceptual tools are insufficient in the face of the loss of the form of life woven by daily care. Nevertheless, the concept of *risk* is clearly no longer appropriate for situations where it is neither calculable nor controllable and where the idea of prevention is meaningless, since the reality is already catastrophic. The very concept of risk then becomes an argument for denial of the agency of care. Not surprisingly, it was Tronto who, in *Le risque ou le care* [16, 17], highlighted the obsolescence of the virilist idea of risk, criticizing Ulrich Beck's classic work, *The Risk Society* [18]. The current crisis is rich in lessons for taking into account the dramatic consequences of the invisibility of women in the North and in the South.

4 Lessons of Intersectionality

The tasks whose importance is recognized today by all are assigned to non-white women (and men too—delivery men, garbage collectors, truck drivers of overwhelmingly foreign origin). The assassination of George Floyd was the occasion of two realizations. First, the incredibly heavy toll paid by African-Americans to COVID killed them in much higher proportions (pending precise data, we are talking about a factor of 2.5). Of course, as in the case of deaths from police brutality, pre-existing conditions are invoked: poverty, obesity, diabetes... However, above all the racist structure of a society that still puts blacks, even today, at the service and at the mercy of whites and has placed them at the front in the struggle of societies against the virus. More than 60% of COVID deaths among caregivers are African-American. They are very numerous, in the USA, in the so-called frontline professions (health, commerce, cleaning, transport, care). They have borne most of the burden of the health crisis and are particularly vulnerable to the looming economic crisis. But they are expendable for the health of the country.

Second, and in many countries, society has taken—quite surprisingly, in fact—a moral position on the global immorality of capitalism: not to sacrifice lives to the economy. Whatever the cost. To whom? The preservation of lives has been the priority, and it was decided to try, first, to cure. This moral choice is also a denial of the lives of those who have been sacrificed for the collective well-being, this collective being in reality those who have remained sheltered and at whose service has been placed the bulk of the work of care provided (care, that is, in the broad sense of caring for the lives of others), here as elsewhere, by the most vulnerable.

5 Conclusion

Democratic societies have thus displayed, with more or less enthusiasm, and respected, with more or less effectiveness, a right to life and a duty to protect populations. We have seen that this protection does not exist for everyone, particularly in France. However, it is totally absent for African-Americans. The death of George Floyd appears to be in line with the fate of African Americans at the time of COVID-19. Black lives matter takes on a new more tragic meaning with the pandemic, and the pandemic might have taught us a lot about entrenched inequalities between people who receive good care and others who are expendable at the service of others.

One can remember the moment of former president Trump's hospitalization for COVID-19. Far from revealing Trump's fragility, his illness showed the inequalities that characterize capitalist societies, and the place of the privileged in the crisis that they most often go through with the help of others, not just with first-class medical treatment but with all they help they receive from drivers, assistants, delivery men to maintenance and home help-personnel, and even caregivers. It shows the extent to which the privileged are making others bear the brunt of the pandemic in general. It was indeed tempting to see in Trump's disease a fair return and proof of the indifference of the virus, which would attack presidents as well as the poorest people. But the numbers tell a different story. In the United States, blacks and Latinos are approximately two to three times more likely than whites to contract COVID-19, three times more likely to die from it. According to the Centers for Disease Control and Prevention, of the 121 children who died from the virus in July in the USA, nearly 80% were Latino or black. Also in France, an INSEE study showed that mortality from COVID was twice as high for people born abroad than for those born in France. A *département* such as Seine-Saint-Denis, with its inhabitants making greater use of public transport to get to work, and working in Paris in sectors such as food, cleaning and delivery, has thus experienced a very high excess mortality rate; in short, these people are working in so-called *essential* care professions, another antiphrase to say that those who perform them in the service of others are negligible in number.

Racial and gender disparities in health are certainly not new, but they take on particular acuteness in a global context where part of humanity has been massively mobilized or vulnerabilized to care for others. Trump alone symbolizes this exploitation. His demented carelessness is, however, only the concretization of the profit system that put him in power, which consists of making others—the most vulnerable—carry the burden of the lives of the privileged. And, this is the last lesson of the COVID.

Core Messages

- The COVID pandemic was a global tragedy, but it also presented a strange pedagogical moment.
- The perspective of care includes an ethical and political ambition, which is not only an active benevolence towards those who are close to us but constitutes *an education* in the perception and in the valorisation of human activities. This is a reminder of the importance of rethinking care and outsourcing or service together.
- The current crisis is rich in lessons for taking into account the dramatic consequences of the invisibility of women in the North and in the South.
- In many countries society has taken a moral position on the global immorality of capitalism: not to sacrifice lives to the economy. Whatever the cost.

References

1. Lovell A, Pandolfo S, Das V, Laugier S (eds) (2013) Face aux désastres. Ithaque, Paris
2. Laugier S (2020) Politics of the ordinary. Care, ethics, and forms of life. Peeters, Louvain
3. Tronto JC, Fisher, B (1990) Toward a feminist theory of caring. In: Abel E, Nelson M (eds) Circles of care. SUNY Press, New York, pp 36–54
4. Tronto JC (2009) Un monde vulnérable, pour une éthique du care. La Découverte, Paris
5. Tronto J (1993) Moral boundaries. A political argument for an ethic of care. Routledge, New York
6. The program: https://www.thelancet.com/journals/lancet/article/PIIS0140-6736(20)30526-2/fulltext
7. Hochschild AR (2000) Global care chains and emotional surplus value. In: Giddens A, Hutton W (eds) On the edge: living with global capitalism. Jonathan Cape, London, pp 130–146
8. Coser LA (1973) The obsolescence of an occupational role. Soc Forces 52(1):31–40
9. Ehrenreich B, Russell Hochschild A (eds) (2002) Global woman: nannies, maids and sex workers in the new economy. Henry Holt, New York
10. Parrenas RS (2002) The care crisis in the Philippines: children and transnational families in the new global economy, in [9], pp 39–54
11. https://www.ipcc.ch/report/srccl
12. Aguilar Revelo L (2009), Manuel de formation sur le genre et le changement climatique. Absoluto, San José (Costa Rica). https://portals.iucn.org/library/node/9395
13. http://www.unesco.org/new/fr/natural-sciences/priority-areas/gender-and-science/cross-cutting-issues/climate-change-and-gender-equality/
14. Hache E (2016) Reclaim, recueil de textes écoféministes. Textes choisis et présentés par Émilie Hache, postface de Catherine Larrère. Éditions Cambourakis, Paris
15. Gupta A, Harrison PJ, Wieslander H, Pielawski N et al (2019) Deep learning in image. Cytometry Rev 95:366–380. https://doi.org/10.1002/cyto.a.23701
16. Tronto JC (2012) Le risque ou le care? Presses universitaires de France, Paris
17. Laugier S, Vallaud Belkacem N (2021) La société des vulnérables. Gallimard, Paris
18. Beck U (1992) Risk society: towards a new modernity. Sage, London

Sandra Laugier is Professor of Philosophy at Université Paris 1 Panthéon Sorbonne, and Senior Fellow of Institut Universitaire de France; Chevalier de la Légion d'Honneur, 2014, Grand prix de philosophie, Académie française, 2022. Translator of most of Cavell's work in French, she is an advisor for the publication of Cavell's Nachlass and the Editor of Here and There (Harvard U. Press, 2022). She is P.I. ERC—Advanced Grant project DEMOSERIES https://www.demoseries.eu.

She has been Visiting Professor at the Cornell School of Criticism and Theory (2023) Toronto University (2022), Boston University (2021, 2019), La Sapienza Roma (2019), Pontifical University Lima (2017), Johns Hopkins University, 2009, 2008, 2011. Visiting Researcher at Max-Planck-Institut für Wissenschaftsgeschichte Berlin, 2014 and 2015, Distinguished Chaire invitée Facultés Saint-Louis, Bruxelles, 2009.

Among her recent publications*Formes de vie* (ed. with E. Ferrarese), CNRS 2018; *Le pouvoir des lien faibles* (ed. with A. Gefen), CNRS 2020, *Politics of the Ordinary, Care, Ethics, Forms of life*, Peeters 2020; *La société des vulnérables* (with N. Vallaud Belkacem), Gallimard 2021.

Chapter 7
Phenomenology of Vulnerability: A Person-Centred Approach

Massimiliano Aragona and Giovanni Stanghellini

> *... by situation we understand changing formations of the original relatedness of person and world, within which talents, qualities, attitudes, constitutions become evident.*
> *Hubertus Tellenbach, Melancholy, 1980*

Abstract The COVID-19 pandemic has caused global stress. However, people reacted differently, some with various forms of distress. Although vulnerable groups were more likely to suffer in this situation, heterogeneity was also reported at this level. We revised the concept of vulnerability in medicine and mental health using schizophrenic symptoms as paradigmatic examples. In our view, studying vulnerability to stressful situations, the following points are fundamental: (1) the individual *emotional* reaction; (2) the personal *meaning* attributed to the events; (3) the way the person concretely *takes position* in front of this (distressing and possibly uncanny) experience; and (4) the way and extent to which the person has *contributed* to the determination and constitution of the stressful situation. Applied to the pandemic situation, we suggest that the reported heterogeneity of reactions depended on the personal way the persons took position in front of the situation they were experiencing. The way they took position depended on the specificities of the situation to be faced but also on their personal way of managing the consequent imbalance of their existential dimensions of anthropological vulnerability.

M. Aragona (✉)
The Italian National Institute for Health, Migration and Poverty (INMP), Rome, Italy
e-mail: massimiliano.aragona@inmp.it

G. Stanghellini
Department of Health Sciences, University of Florence, Florence, Italy

"Diego Portales" University, Santiago, Chile

G. Stanghellini
e-mail: giovanni.stanghellini@unifi.it

M. Aragona
The Roman Circle of Psychopathology, Rome, Italy

S. Achella and C. Marazia (eds.), *Vulnerabilities*, Integrated Science 18,
https://doi.org/10.1007/978-3-031-39378-5_7

Keywords Phenomenological psychopathology · Vulnerability · Schizophrenia · Philosophy of psychiatry · COVID-19

1 Introduction

The COVID-19 pandemic, with related lockdowns and other restrictive social measures, represented a global event causing stress all over the world. Several authors have emphasized the possible traumatic nature of this experience and its possible psychopathological output in terms of mental distress. Although it is difficult to measure the real impact of the pandemic on mental health because studies evaluating the same persons before and during/after lockdowns are rare, a recent systematic review and meta-analysis is available [1]. In countertendency, it suggests that (a) soon after the outbreak of the COVID-19 pandemic, there was a small increase in mental health symptoms, but they decreased at prepandemic levels by mid-2020; and (b) change in mental health was highly variable across samples. These results are reinforced by another systematic review stating that in those who had COVID-19 infection, the long-term prevalence of anxiety, depression, PTSD, and sleep disturbances is comparable to that found in the general population [2]. In sum, it seems that the long-term impact of the pandemic on the mental health of the general population is not as alarming as expected, suggesting that not everyone, but only those persons in vulnerable conditions, may have significantly suffered. Indeed, in vulnerable groups (people in poor socioeconomic conditions, homeless individuals, migrant workers and asylum seekers/refugees), accessing mental health services was very difficult during the pandemic period, with problems also in treatment adherence [3]. However, even in such 'vulnerable' groups already in treatment for mental health issues, the personal experience of the pandemic was heterogeneous, i.e., some had worsened during the lockdown, while others were stable or improved [4]. A lesson we can learn from it is that although all patients were labelled 'vulnerable', one thing is to expect (in general) a negative reaction to a living difficulty or a stressful event because the person belongs to a 'vulnerable' group; another issue is the actual reaction of that single person in that unique life situation.

Recently, the debate has focused on the question of whether vulnerability refers to a group of persons (categorical approach) or to the consequences of situations and context (contextual approach) [5, 6]. This is a move trying to overcome the limitations of the categorical approach, focusing on the specific set of conditions faced by the person, with their own potential for stress induction. It is a step forwards because in this view, it is not the person who is vulnerable *in se* but the unique dynamic set of interrelated conditions that cooperate in making that situation so stressful for the person that the consequent reaction is expected and understandable.

Nevertheless, both views have in common a similar key point: i.e., in either case, the concept of vulnerability is grounded on probabilistic reasoning about risk factors. In other words, both views are based on a mechanistic approach:, i.e., provided the existence of some risk factors acting as stressors, the computation of their force in disrupting 'normal' functioning can be measured, and their effect can be predicted. In the categorical approach, the question is "how much does the sum of the factors make that person vulnerable because *he/she belongs to that group*?", while in the contextual approach, it is "how much is the sum of the factors acting on that person *in that specific situation*?".

It is not to us to deny the relevance of these concepts and the usefulness of these studies in assessing the probabilities of negative outputs in populations exposed to stressful events. Indeed, in public health, this information is fundamental for the correct planning of health facilities and activities in territories. We just note that such mechanistic explanations do not expound psychopathological reactions *without residuum*, i.e., they are insufficient to tell why some persons with similar vulnerability profiles and exposed to similar vulnerable situations reacted differently.

Our point is that this is because in these models, the *experiential* component of the *personal* reaction is neglected. That is, in our view, it is fundamental (1) the individual *emotional* reaction, (2) the personal *meaning* attributed to the events, and (3) the way the person concretely *takes position* in front of this (distressing and possibly uncanny) experience. Last but not least, it is also important to acknowledge (4) the way and extent to which the person has *contributed* to the determination and constitution of the stressful situation [7, 8]. Mechanistic, probabilistic computation can assess the likelihood of a reaction but cannot recognize the specific experiential features of a single reaction. The latter depends not only on general probabilities but also on concrete ways the person has to handle the situation, which largely relies on the meaning, emotional relevance and value he/she ascribes to it.

In this chapter, we use psychopathological research as a case in point to show how the concept of vulnerability has been reshaped in recent years, moving from a categorical/mechanistic view (vulnerability as a genetic diathesis, an inherited weakness of/in the body that makes it prone to dysfunction under stress) to a phenomenological one (a dialectical interaction between the person and her/his basic uncanny experience of the situation). Implications for research on COVID-19 and mental health are drawn.

2 The Diathesis-Stress Model

This classical model explains disorders as the result of an interaction between a predispositional vulnerability and the stress deriving from life experiences. Such vulnerability is called diathesis from the Greek διάθεσις, meaning 'disposition', which acquired the sense of 'bodily disposition' already in Aristotle. Although in theory both diathesis and stress could be psychological, in the advancement of research, it was their biological/somatogenic interpretation that tended to prevail, particularly in schizophrenia research. Classically, the diathesis-stress model was interpreted as a model suggesting how biological or genetic traits (the diathesis) interact with environmental factors (the stressors) to produce disorders such as schizophrenia. Even if schizophrenia is not the only disorder interpreted in this way, it is a paradigmatic case and will be used in this chapter as a good example to show how models of psychopathological vulnerability are shaped.

In the classical diathesis-stress model, when the social-relational stress exceeds the threshold represented by the biologically settled capacity of resistance, the disorder onsets. However, with time, the stress component also tended to be interpreted as biological. For example, in a frequently quoted article, it was suggested that stressors should be defined as events or experiences that jeopardize homeostasis through an imbalance of neuroendocrine and autonomic nervous system reactions. In the case of schizophrenia, it was suggested that constitutional vulnerability involving an abnormality in the dopamine neurotransmission system and stress-induced changes in hypothalamic–pituitary–adrenal (HPA) function interacted to produce schizophrenic symptoms [9]. Recently, revisited, the model proposes a "more nuanced understanding" of hypothalamic–pituitary–adrenal axis function that should include the role of neurodevelopmental, epigenetic, neurotransmitter, and inflammatory processes, as well as brain structure and function [10]. In sum, with time, the model increases in complexity but basically remains confined to the biological realm of explanation. It seems that within this model, there is no space for human meaningful activity in shaping psychopathological reactions, provided that the challenges for future research are "identifying the genetic and environmental determinants of vulnerability and the neurobiological pathways mediating the effects of exposure to adversity" [11].

3 Schizophrenia: From Patent Symptoms to Core Phenomena

Before discussing the phenomenological concept of vulnerability in the development of schizophrenic symptoms, it is useful to show the progress made in the last thirty years of psychopathological research on this issue.

Currently, the diagnostic criteria for schizophrenia that are used to communicate among clinicians are those listed in textbooks such as the DSM-5 or the ICD-11. In both cases, the reader finds a list of symptoms and a numerical threshold stating the lowest number of criteria of the list that are necessary for the diagnosis. For example, in the case of DSM-5, the first criterion requires that at least two of the following symptoms were present for a significant portion of time during a 1-month period (or less if treated). The symptoms are: delusions, hallucinations or disorganized speech (at least one of them is necessary), grossly disorganized or catatonic behaviour, and negative symptoms (e.g., lack of volition or reduced emotional expression).

Now, setting aside the (all but irrelevant) question of what is meant by delusion, hallucination, etc., two main problems arise from this definition. First, there are other disorders that present with symptoms such as delusions and hallucinations, that last for one month or more and that are not schizophrenic—, i.e., the criterion is satisfied by various disorders, not only by schizophrenia. Second, there are schizophrenic patients without delusions, hallucinations or disorganized speech among their symptoms, i.e., the criterion is not satisfied, although they are viewed as schizophrenic by clinicians. To be rigorous these patients should not be diagnosed as having schizophrenia but experienced clinicians are well aware of their existence, and they fit other more classical criteria like those settled by Kraepelin or Bleuler, i.e., the fathers of this diagnostic concept.

In sum, the official symptoms of schizophrenia are not as specific of this diagnosis as necessary for clinical management, so clinicians found it necessary to go beyond them, looking for more fundamental, essential phenomena. Said differently, they moved from "nosographical organizers" to "psychopathological organizers", described as schemes of comprehension that organize psychopathological experiences around unitary cores of meaningfulness [12].

In this vein, psychopathological research on schizophrenia put aside the conventional and never proven biological models proposing a link between patent symptoms and supposed neurobiological dysfunctions and redirected its focus on core schizophrenic phenomena that can be at the basis of the development of symptoms.

Psychopathological research following this direction emphasized at least three points of interest.

First, subtle alterations of perception, disturbances of thought processing and/or aberrations from normal cognitive processes and functions, independent of though thought content (i.e., mainly formal alterations), have been described. These are considered basic phenomena arising in a preclinical stage (sometimes many years before the onset of manifest symptoms), and it is believed that the latter appear only when the person's attempts to cope with the basic symptoms fail, reaching the limit of her/his ability to compensate [13].

Second, the existence of a basic alteration in the flow of the common experience, such that experience loses its obvious, commonsensical quality, which characterizes those form of schizophrenia that do not develop the DSM-5 symptoms and thus do not satisfy the number of criteria necessary to have the DSM diagnosis. In the XX century, many of these forms could have been diagnosed as cases of hebephrenic or simplex schizophrenia and were finally defined as *subapophanic* (meaning subdelusional) schizophrenia by Blankenburg [14].

Third, the general affective-atmospheric *colour* of the experience immediately before the onset of delusion, which the classical authors called *Whanstimmung*. It is in this brief but decisive stage that delusional ideas and other psychotic phenomena arise and take their form. Accordingly, Ballerini and others [15] stressed that schizophrenic delusion is not truly *primary*; rather, it is the active effort of the person to give sense to the perplexing uncanny atmosphere experienced in the predelusional phase.

In summary, psychopathological research recognized the limitation of those models linking DSM schizophrenic symptoms and possible underlying neurobiological dysfunctions and looked for more subtle phenomena that could underlie DSM symptoms. Basic symptoms, loss of natural evidence and predelusional emotional/atmospheric disturbances are examples of this. Here, we suggest that this new emphasis on core phenomena also changes the concept of vulnerability of schizophrenia, as we will show in the next section.

4 From Biological Diathesis to Core Experiences and Hermeneutic Position-Taking

In psychiatry, the classical diathesis-stress model conceives a biological vulnerability on which the stressors act. The model is often interpreted quantitatively: a "normal" amount of stress acting on persons with good resilience is handled without symptoms; an extremely high amount of stress can produce symptoms even in persons with good resilience; and a "normal" (or not extreme) amount of stress acting on vulnerable persons with low resilience can lead to symptoms. Stressors are often interpreted as social variables, but as seen above, their biological effect (e.g., in terms of imbalance of stress hormones such as cortisol) is also considered.

In schizophrenia, the mental condition we choose as an example, according to the classical diathesis-stress model, there would be a vulnerable individual (usually it is a genetic vulnerability, leading to neurobiological "endophenotypes"[1]). Such vulnerable individuals find it difficult to manage the stressful inputs of adolescence (both biological, e.g., hormonal changes, and relational), there is a sort of breakdown, and finally, symptoms arise. In other words, in this model, schizophrenic symptoms arise in a biologically vulnerable person under stress as an output of a stress-induced neurobiological dysfunction. However, it was difficult to discover a real link between such symptoms and the supposed neurobiological vulnerability. Consequently, as shown in the previous section, psychopathological research moved from patent DSM symptoms to subtle core phenomena as possible alternative proxies for the supposed neurobiological dysfunction.

Here, we stress another step that sometimes has passed unnoticed. Our claim is that while moving from DSM symptoms to core phenomena, researchers also shifted the emphasis from neurobiological explanation to living experience. Indeed, the importance of core phenomena is not only that they are closer to the neurobiological level (which is often stated but remains an assumption) but also that they are basic *experiences*. Core phenomena bring to the fore something different in the person's experience, something deserving an active "position-taking". By *position-taking* [16], the philosopher Husserl meant a class of intentional acts that are not merely restricted to being conscious of a given object but entail an active and, to a certain degree, free orientation toward that object. In other words, when we encounter other persons, things or states of affairs in our everyday lives, we do not passively face them in a pregiven manner. Rather, we (at least partly) actively orient ourselves towards a certain situation so that we become acquainted with it perceptually, affectively, cognitively, and the like [17]. As a result of this taking of position, the situation is then constituted as such in its meaningful structure and drives explicit actions. Applied to psychopathological research, a phenomenologically informed practice of this kind involves not only an accurate assessment of signs and symptoms but also a recognition of the person's position-taking, including the dynamic effects of such position-taking on the manifestation and course of psychopathological phenomena. In the case of

[1] Endophenotypes are usually described as neurobiological altered functions that are in the middle of a causal chain between genetic contributions to a disorder and its diagnosable mental symptoms.

schizophrenia, in this view, full-blown DSM symptoms (e.g., delusions, hallucinations, etc.) emerge through a complex and dynamic interaction between anomalous changes in implicit aspects of experience (i.e., the core phenomena described in the previous section) and the person's various self-interpreting attempts enacted in the face of those uncanny experiential changes. Hence, what we call schizophrenic "symptoms" are not single, well-defined and stand-alone entities that can be mapped straightforwardly onto an altered neurobiological network. Rather, they are the final expression of the person's efforts at compensating, making sense and adapting to the challenges of meaninglessness, hopelessness, passivity and despair that often characterize his/her basic distressing experiences. Schizophrenic symptoms—both "positive" and "negative" symptoms—thus derive from an attempt at position-taking with respect to these experiences [16].

As a result of this reinterpretation of the meaning of schizophrenic symptoms, we can derive three claims that may also be relevant for a general discussion on vulnerability.

First, core experiences not only represent a possible biologically grounded vulnerability but also derive their importance from being *experiential vulnerabilities*, i.e., lived experiences so uncanny that the person has to react by taking a position to manage them in some way.

Second, the way two persons take positions in front of similar life events may be *very different* depending on their personal stance. Different persons can experience the same life event in different ways. The difference between 'objective event' and 'subjective experience' has long been established [18]. The same event can be experienced in different ways by different persons, according to their personal vulnerability [19] or resilience. As a consequence, in some cases, one person can develop symptoms, and the other person can develop other forms of reaction, including nonpathological reactions.

Third, mental symptoms are the final output of a *personal hermeneutic* (i.e., interpretative, sense-making) *process* starting from core experiences. In this sense, anomalous world- and self-experiences represent the rugged terrain within which symptoms such as delusions and hallucinations may develop through a reactive or interactive process that leads from deep to surface phenomena. Full-blown symptoms (such as delusions and hallucinations) are *emergent* phenomena in two senses: first, they emerge under stress as experiential manifestations of an underlying personal vulnerability, even if it can be sometimes difficult to pinpoint the stressful situation; second, they emerge out of the interpretation the person makes her anomalous and uncanny experiences as they arise in the course of the stressful situation itself. All this can be summarized as follows:

- The person, with her/his set of habits, emotions, needs, desires, beliefs and values, contributes to the genesis/constitution of her/his stressful situation
- The stressful situation is not a given "objective" life event *in se*, but rather it is "stressful" *for* that given person.
- In a stressful situation, the person develops a given (set of) emotion(s) (e.g., shame, anxiety, fear, etc.) which characterizes her/his feelings about the situation

- The person interprets the situation according to her/his mindframe (including preexisting beliefs, values, etc.) and (often involuntarily and unconsciously) attributes a personal meaning to it.

5 From Neurobiology and Social Risk Factors to Anthropological Vulnerability

We started from the claim that research on vulnerability is usually based on mechanistic models in which both neurobiological mechanisms and social issues are quantified and treated statistically as risk factors interacting in a cause-effects chain.

In the example of schizophrenia, the mainstream vulnerability model claims that biologically vulnerable individuals are unable to manage an excessive degree of stressful inputs so that normal functioning goes into crisis and symptoms arise. Accordingly, mental symptoms are the output of a basically vulnerable neurobiological mechanism impaired by too much stress, i.e., the final effect of a dysfunctional mechanism.

In the case of vulnerability models applied to the stress generated by the COVID-19 pandemic, the stress is generalized, but its effect on individuals differs depending on factors such as the social vulnerabilities of the individual (e.g., job loss) and personal vulnerability (e.g., preexisting mental disorders, usually interpreted as biological vulnerability).

In both cases, the interpretation is in line with a diathesis-stress model looking for quantitative factors interacting in a mechanical chain.

We do not deny the importance of studies on biological and social vulnerabilities in the measurement of risk factors for the development of mental distress in various populations, in general as well as in the specific case of the COVID-19 pandemic. Our claim is that these quantitative studies should be complemented by qualitative studies exploring the lived experience of persons and their active role in shaping mental distress. In other words, the person has an active and idiosyncratic reaction to the experiential novelty introduced by the new situation, and this very reaction is part of the construction of possible manifest symptoms of mental distress.

We selected the case of schizophrenia as a paradigmatic example because it is a psychiatric disorder often considered a "nonunderstandable" output of strictly biological origin and interpreted in the lines of the diathesis-stress model. If we were right in our analysis of schizophrenic vulnerability, showing that schizophrenic full-blown symptoms are secondary constructions of a personal hermeneutic (partially implicit) reaction to the experience of core phenomena [20], a fortiori we shall be right in more "understandable" cases such as psychological distress in reaction to the experience of the pandemic.

Before concluding, we use this last part of the chapter to expound the notion of anthropological vulnerability as emerged in schizophrenia studies. Core experiential structures of the life-world of persons with schizophrenia, such as subtle disorders of selfhood and world-experience, are usually neglected in mainstream research, diagnostic manuals and structured interviews, mainly because of their elusive qualities [21]. However, the emergence of patent schizophrenic symptoms would be totally incomprehensible without the phenomenological reconstruction of the active role of the patient in taking position in front of her/his basic anomalous experiences. In other words, in addition to the genetic vulnerability postulated in the diathesis-stress model, there is a psychopathological vulnerability to schizophrenia that is based on structural changes in subjectivity. Accordingly, the core or generative abnormality in schizophrenia is supposedly a particular kind of disturbance of the prereflexive self, which is at the root of basal sensations of alienation [21]. Such basic uncanny experiences, which we described above as "core phenomena", are those from which the hermeneutic reaction starts leading to the final DSM symptoms. Conceptualizing schizophrenic vulnerability in relation to the notion of self-disorder means that in schizophrenia, self-disorder is not a definite and constant disturbance. Rather, the claim is that disordered selfhood is a vulnerable trait that is constantly threatened and unstable [22].

Now that we have completed the shift from a mechanical interpretation of vulnerability to an experiential vulnerability in which the illness is the product of a hermeneutic activity of position-taking in front of basic unstable experiential disturbances, it remains to enlarge the concept at the anthropological level. Ludwig Binswanger, the father of phenomenological psychopathology, is credited for having reframed mental disorders as particular idiosyncratic ways of being, instead of medical diseases. In his view, mental symptoms are the expression of an existential failure rather than the effect of a dysfunctional mechanism. Using Heidegger's philosophy at the *ontic* level (i.e., applied to the particular life history of patients), Binswanger imported into psychiatry the analysis of the *existentialia* (*Existenzialen*) as those anthropological general categories indicating the fundamental ways of being in the world. It was stressed that categories of thought such as the interpersonal "being-together" (*mit-Dasein*), the view of the objects as something defined by their use for human beings, the lived-space and the lived-time, "are fundamental anthropological contributions by Heidegger, strongly inspiring generations of phenomenologically oriented psychopathologists" [23].

Drawing on these concepts and extending the discussion to the contribution of other anthropologically oriented psychopathologists (e.g., Minkowski, Straus, von Gebsattel), one of us [24] proposed the project of an anthropological psychiatry redirecting abnormal modes of being in the world towards the categories of human existence. In the case of vulnerability to the onset of schizophrenic symptoms, pathological developments are viewed in light of the unbalance between anthropological dimensions such as authenticity–inauthenticity, proportion–disproportion, eurhythmy–arrhythmia, and activity–passivity. In this anthropological view, schizophrenic symptoms are the result of a failure to maintain a comfortable proportion between these poles. This leads to alienation from one's experiential base and from the dimension of intersubjectivity; to the disappearance of rhythmic, harmonious pulsations and the diachronic relation between the diastolic and systolic phases of life; to the incapacity to let things be without questioning the obvious; and to the experience of becoming passive in front of the world with the impossibility of recounting this experience [24].

Similarly, we can ask if the generalized stress due to the pandemic has impacted differently depending on the anthropological axes unbalanced in the life experience of single persons.

For example, how has forced social distancing changed interpersonal relationships? Normally, trust and suspicion in everyday relationships are balanced depending on familiarity and other social cues. During the pandemic, everybody was a potential spreader of the infection, greeting each other by embracing and kissing (which is usual in many cultures) was forbidden, isolation was increased, etc. It is likely that these changes have impacted individuals differently. In our opinion, this was not because they were included in a given category of vulnerability or in a diagnostic group but because different persons differently balance the dialectical relationship between trust and suspicion. Accordingly, those with a preexisting tendency toward suspiciousness (not necessarily pathological) may have reacted with an increased sensitivity to threatening messages, hence increasing social isolation, fear of contagion, anxious check of the news about the pandemic, compulsive cleaning, etc. In contrast, others with a preexisting tendency toward social isolation have described an experience of relief because they weren't the weird ones anymore, everyone was isolated so that they felt to be more "normal" than before.

The examples could continue, but these are enough to stress our point: in the same condition, people reacted differently depending on their idiosyncratic way of giving meaning and managing the situation. In our examples, the pandemic situation altered an anthropological equilibrium, and the way the person experienced and reacted to this disequilibrium was part of the process of position-taking leading to the development of new forms of adaptation, distressful or not depending on the case.

The examples above are related to the way-of-being with others and with objects. Other phenomenological *existentialia* can be addressed as well. For example, the personal experience of the flow of time (the *lived time*) was disrupted in the experience of many during the pandemic, and the rhythmicity as well (think for example to the delay in going to sleep of many adolescents). Moreover, the way the space (the *lived space*) was experienced, particularly during the lockdown, the experience and the care of the body (some remaining in pyjamas or tracksuit for days, beauty salons and hairdressers banned, eating habits disrupted, etc.), the dialectics between rules and freedom, that between activity and passivity, are all good examples of existential dimensions whose equilibrium was perturbed during the pandemic. Talking with people and taking care of their personal experience, it would be easy to show many other examples of personal, often implicit, hermeneutic position-taking in front of the general pandemic situation.

Unfortunately, among 21,110 articles addressing COVID-19 and mental health, only 37 (0.17%) use the term "phenomenology" (PubMed search performed on Jan 4th, 2023), suggesting that there is an informative gap at this level and consequently a strong need to support qualitative research about the *lived experience* of people during the pandemic.

6 Conclusions

In this chapter, we discussed the concept of vulnerability and showed that categorical and contextual/situational vulnerability share a quantitative approach that is useful for evaluating risk factors and measuring general probabilities but is unable to account for personal reactions. We argued that a qualitative, phenomenological approach is needed to complement available quantitative information and make it understandable the different ways people react in the face of similar situations.

The example of schizophrenia was used to show in detail how research progressed from a mechanistic, mainly biologically oriented model of diathesis-stress interplay to phenomenological models considering vulnerability as a precarious equilibrium between opposite poles of anthropological dimensions. In such models, the development of psychiatric symptoms is not the output of broken mechanisms but the way the person takes position in front of the uncanny experiences that unbalance such a precarious equilibrium. Accordingly, symptoms are not effects but active constructions, ways to try to cope with uncanny experiences. In some cases, this movement is fruitful (for example, in cases of "post-traumatic growth", when personal growth is reached after passing through a period of traumatically elicited distress); in other cases, it is not. In the case of schizophrenia, many symptoms (delusions, hallucinations, social withdrawal) can be seen as efforts to make sense of radically uncanny and nonunderstandable experiences. Ways to manage the situation that reduce the angst and perplexity of the predelusional stage but at the cost of alienation from the common world.

Finally, we briefly considered whether the insight coming from psychopathological research could also be fruitfully used in studies addressing the effects of the pandemic situation on the mental health of the population. Even in this case, we stated that quantitative studies on vulnerable populations are useful, but they do not explain why different people (even if all included in the same vulnerable group) reacted differently to an apparently similar stressful situation such as the global pandemic and the social restrictive measures used by governments to mitigate its effects. In the final part, we briefly sketched a few examples of interplay *at the experiential level* of the personal meaning attributed to the different situations and the anthropological vulnerability represented by precarious balances between different poles of existential dimensions.

Considering the scarcity of phenomenological qualitative research on these issues, we suggest that further research should be implemented in this field. This would be particularly relevant because it would present a more reliable picture of the situation, avoiding generalizations.

We all suffered from the pandemic, but we did not all suffer to the same extent.

Vulnerable groups suffered more from the pandemic, but their members did not all suffer to the same extent.

During the pandemic, people facing particularly vulnerable situations suffered, but they did not all suffer to the same extent and in the same way.

During the pandemic, every person gave his/her own meaning and significance to what was going on, and they reacted differently depending on their own personal way to take-position in front of the situation they were experiencing. The way they took position depended on the specificities of the situation to be faced, of course. However, also from their personal way to manage the consequent imbalance of their existential dimensions of anthropological vulnerability.

Core Messages

- Current categorical and contextual/situational concepts of vulnerability share a quantitative approach that is unable to account for personal meaningful reactions.
- Based on psychopathological research, we described phenomenological models considering vulnerability as a precarious equilibrium between opposite poles of anthropological dimensions.
- In these qualitative models, psychiatric symptoms are personal ways to take position in front of distressing experiences.
- Trying to cope with uncanny experiences, symptoms usually reduce the distress but also life possibilities (e.g., the self-confinement in restricted life-situations to prevent interpersonal distress). However, the same dynamics can be responsible of positive effects (e.g., the personal growth triggered by the experience of self-strength in being able to manage difficult situations).
- Considered from this point of view, the personal reactions to the experience of the COVID-19 pandemic were also related to the specific existential unbalance induced by the situation and the different ways the individuals managed it.

References

1. Robinson E, Sutin AR, Daly M, Jones A (2022) A systematic review and meta-analysis of longitudinal cohort studies comparing mental health before versus during the COVID-19 pandemic in 2020. J Affect Disord 296:567–576
2. Bourmistrova NW, Solomon T, Braude P, Strawbridge R, Carter B (2022) Long-term effects of COVID-19 on mental health: a systematic review. J Affect Disord 299:118–125
3. Aragona M, Barbato A, Cavani A, Costanzo G, Mirisola C (2020) Negative impacts of COVID-19 lockdown on mental health service access and follow-up adherence for immigrants and individuals in socioeconomic difficulties. Public Health 186:52–56
4. Aragona M, Tumiati MC, Ferrari F, Viale S, Nicolella G, Barbato A, Cavani A, Costanzo G, Mirisola C (2022) Psychopathological effects of the Coronavirus (Sars-CoV-2) imposed lockdown on vulnerable patients in treatment in a mental health outpatient department for migrants and individuals in poor socioeconomic conditions. Int J Soc Psychiatry 68:203–209
5. Gordon BG (2020) Vulnerability in research: basic ethical concepts and general approach to review. Ochsner Journal 20:34–38
6. Ferreira JBB, Santos LLD, Ribeiro LC, Rodrigues Fracon BR, Wong S (2021) Vulnerability and primary health care: an integrative literature review. J Primary Care Commun Health 12:21501327211049704
7. Tellenbach H (1980) Melancholy: history of the problem, endogeneity, typology, pathogenesis, clinical considerations. Duquesne University Press, Duquesne

8. Castignoli G (2019) The situative and the reactive. Dial Philos Mental Neuro Sci 12:69–74
9. Walker EF, Diforio D (1997) Schizophrenia: a neural diathesis-stress model. Psychol Rev 104:667–685
10. Pruessner M, Cullen AE, Aas M, Walker EF (2017) The neural diathesis-stress model of schizophrenia revisited: an update on recent findings considering illness stage and neurobiological and methodological complexities. Neurosci Biobehav Rev 73:191–218
11. Walker E (2022) The permeable boundary between biological and psychological processes in the origins of psychosis. Psychiatry Res 317:114853
12. Rossi Monti M, Stanghellini G (1996) Psychopathology: an edgeless razor? Compr Psychiatry 37:196–204
13. Huber G, Gross G (1989) The concept of basic symptoms in schizophrenic and schizoaffective psychoses. Recenti Prog Med 80:646–652
14. Blankenburg W (1971) Der Verlust der natürlichen Selbstverständlichkeit. Ein Beitrag zur Psychopathologie symptomarmer Schizophrenien. Enke, Stuttgart
15. Ballerini A (1999) Esiste una specificità del delirio schizofrenico. In: Rossi Monti M, Stanghellini G (eds) Psicopatologia della schizofrenia. Prospettive metodologiche e cliniche. Cortina, Milano
16. Stanghellini G, Aragona M, Gilardi L, Ritunnano R (2022) The person's position-taking in the shaping of schizophrenic phenomena. Philos Psychol, pp 1–26
17. Arango A (2014) Husserl's concept of position-taking and second nature. Phenomenol Mind 6:168–176
18. Binswanger L (1928) Lebensfunktion und innere Lebensgeschichte. Monatsschr Psychiatr Neurol 68:52–79
19. Kretschmer E (1918) Der Sensitive Beziehungswahn. Ein Beitrag zur Paranoiafrage und zur Psychiatrischen Charakterlehre. Springer, Berlin
20. Stanghellini G (2019) The PHD method for psychotherapy: integrating phenomenology, hermeneutics, and psychodynamics. Psychopathology 52:75–84
21. Stanghellini G, Fusar-Poli P (2012) The vulnerability to schizophrenia mainstream research paradigms and phenomenological directions. Curr Pharm Des 18:338–345
22. Stanghellini G, Rosfort R (2015) Disordered selves or persons with schizophrenia? Curr Opin Psychiatry 28:256–263
23. Stanghellini G, Aragona M (2016) Phenomenological psychopathology: toward a person-centered hermeneutic approach in the clinical encounter. In: Stanghellini G, Aragona M (eds) An experiential approach to psychopathology: what is it like to suffer from a mental disorder? Springer, Cham, pp 1–43
24. Stanghellini G (2018) Metaphors of pathology. Int Rev Psychiatry 30:140–152

Massimiliano Aragona, psychiatrist, psychotherapist and philosopher, is the clinical director of the Mental Health Unit at the Italian National Institute for Health, Migration and Poverty (INMP).

Founder of the Roman Circle of Psychopathology and editor-in-chief of the international open-access journal Dialogues in Philosophy, Mental and Neuro Sciences. He chairs the Philosophy & Psychiatry section of the European Psychiatric Association.

Author of books and articles on several issues including clinical psychopathology, epistemology of psychiatry, phenomenological psychopathology, history of psychiatry and psychology, psychopathology of migrations and post-traumatic stress reactions, psychopharmacology.

Giovanni Stanghellini, M.D. and Dr. Phil. *honoris causa*, psychiatrist, is professor of Dynamic Psychology at Florence University (Italy) and Professor Adjuncto "D. Portales" University (Santiago, Chile). He chairs the *International Network for Philosophy and Psychiatry* (with K.W.M. Fulford and J.Z. Sadler). He authored more than 160 Scopus publications on the clinical phenomenology of schizophrenia, major depression, eating disorders and on the phenomenological basis of psychotherapeutic care. Among his books in English*Nature and Narrative* (co-edited with K.W.M. Fulford, K. Morris and J.Z. Sadler, Oxford University Press 2003), *Disembodied Spirits and Deanimated Bodies. The Psychopathology of Common Sense* (OUP 2004)*Emotions and Personhood* (with R. Rosfort, OUP 2013), *One Hundred Years of Karl Jaspers' General Psychopathology* (co-edited with T. Fuchs, OUP 2013), *Oxford Handbook of Philosophy and Psychiatry* (co-edited with K.W.M. Fulford et al., OUP 2013), *The Therapeutic Interview in Mental Health. A Values-Based and Person-Centered Approach* (with M. Mancini, Cambridge University Press, 2017), *Oxford Handbook of Phenomenological Psychopathology* (co-edited with M.R. Broome et al., OUP 2019), *International Perspectives in Values-Based Mental Health Practice. Case Studies and Commentaries* (co-edited with D. Stoyanov, K.W.M. Fulford et al., Springer, 2020) and *Time and Body* (co-edited with C. Tewes, Cambridge University Press, 2020).

Chapter 8
(In)Visibility of Children and Their Psychosocial Vulnerability–The Narrowed Discourse on Children in the First Year of the Pandemic in Germany

Maria Griemmert and Anne Oommen-Halbach

For many children, I am convinced, something will remain of this feeling of being at the mercy of others, of being locked up. As a society, we have already done a kind of triage: We have decided that the lives of the elderly and the at-risk groups are more important than the mental well-being of the children […].
Daniel Kehlmann, Interview, May 2020
Original opening quote: "Bei vielen Kindern, davon bin ich überzeugt, wird von diesem Gefühl des Ausgeliefertseins, des Eingesperrtseins etwas bleiben. Gesamtgesellschaftlich haben wir also bereits eine Art Triage durchgeführt: Wir haben entschieden, dass das Leben der Älteren und der Risikogruppen wichtiger ist als das seelische Wohlbefinden der Kinder […]" [1].

Abstract German children and young people took on an ambivalent role in the pandemic: Largely banished and invisible in the public sphere during the so-called *lockdowns*, their media representation followed abbreviated narratives that considered children primarily responsible for the health of specific risk groups. Their own claims to protection and participation, and their psychosocial vulnerability, which was particularly impaired by the measures of social distancing, initially remained invisible. This paper examines the extent to which child and youth specific issues

M. Griemmert (✉) · A. Oommen-Halbach
Department of the History, Philosophy and Ethics of Medicine, Medical Faculty, Heinrich-Heine-University Düsseldorf, Düsseldorf, Germany
e-mail: griemmm@hhu.de

A. Oommen-Halbach
e-mail: anne.oommen-halbach@hhu.de

S. Achella and C. Marazia (eds.), *Vulnerabilities*, Integrated Science 18,
https://doi.org/10.1007/978-3-031-39378-5_8

were present in German media discourse in the first year of the COVID-19 pandemic, and which issues remained blind spots.

Keywords Children · Pandemic · Schools · Media discourse · Participation · Well-being

1 Introduction

In the drastic words of the opening quote, the German-Austrian writer Daniel Kehlmann reflected at an early stage on the hierarchisation of societal risk and vulnerability assessments redefined in light of the pandemic [1]. With his wake-up call that children and adolescents would be harmed by the particularly far-reaching infection control measures affecting them, he called into question the German coronavirus policy – which until then had been considered a success story [2] – with regard to the well-being of the youngest members of society. While this view was mainly met with criticism and incomprehension in May 2020, Kehlmann's prediction that lasting negative effects would accompany longer-term isolation and school closures for many children and adolescents is, three years later, part of post-pandemic general public knowledge and scientific mainstream: from today's perspective, many psychosocial burdens of children are considered to be side effects of the various demands of social distancing during the coronavirus pandemic[1]. This is also emphasised by Sabine Andresen when she states that the lack of opportunity to leave a stressful situation (voice, choice, exit) during lockdowns contributed to the increased vulnerability of children and young people [6, p. 1]. In retrospect, the question arises of how it happened that the psychosocial needs of children were so disregarded in the public debate of those days?

This paper examines to what extent child- and youth-specific topics were present in the German media discourse during the "hot phase" of the COVID-19 pandemic between spring 2020 and early summer 2021: What was the focus of the media coverage of topics concerning children, and how were they represented in the press? In addition, which subjects relating to children were, on the contrary, invisible, or at least overshadowed?

[1] Studies of psychosocial health such as COPSY [3] point to negative psychosocial effects of social distancing and school closures on young people, as do the observations of educational researchers and pedagogues. Compare, for example [4, 5].

2 Being a Child in the First Year of the Pandemic

Children and adolescents are usually considered to be more vulnerable than adults. Their vulnerability is based on the fact that minors are usually (still) dependent on adult support to meet their basic life needs, such as food, clothing and housing, but also education, socialisation and social bonding through caregivers. However, the characteristic of child vulnerability is its developmental dynamic (and its individual component): the more a child's physical and mental growth progresses, which results in an increase in its capacity for autonomy, the less it requires support from adult caregivers.

Vulnerability can be described as an "umbrella term", which conceptually encompasses various phenomena from different domains [7, p. 185]. One of these domains is a biological one: children's growing bodies are more sensitive to multiple physical influences, such as pathogens or medications. From a medical ethics perspective, children and adolescents are therefore subject to special protection, especially in medical research: research projects involving minors, e.g. have to fulfil particularly high ethical standards [8].

Regarding the threat of the COVID-19 pandemic for children, there was an early all-clear concerning this "biological" vulnerability of children: as early as 2 March 2020, and thus even before the SARS-CoV-2 outbreak was declared a pandemic by the World Health Organisation, the Commission for Infectious Diseases and Vaccination Issues of the German Academy for Paediatrics and Adolescent Medicine (DAKJ) had pointed out that children were less susceptible and physically at risk and therefore "less vulnerable" than elderly or chronically ill people [9].

As a result, minors were no longer considered a group in need of special protection in the immediate threat of the pandemic. Furthermore, children were often no longer included when "vulnerable" or "risk groups" were mentioned [10, pp. 5–6], since "vulnerability", in light of the pandemic, was largely equated with the specific vulnerability to the pathogenicity of the new coronavirus.

2.1 *Invisibility in Public Space*

In the first so-called *lockdown* in spring 2020, which lasted from mid-March to early May for approximately 7 weeks, children and young people in Germany were completely excluded from public spaces. In addition to the closure of schools and day-care centres, sports clubs, etc., this was also ensured by restrictions formulated exclusively for children, e.g. the closure of playgrounds, bans on entering public buildings, orders not to go shopping in family groups, etc. [11].

According to an OECD survey, classes for the approximately eleven million pupils in Germany were disrupted on average more than 180 days between the beginning of the coronavirus pandemic and the end of the school closures in spring 2021. While many countries limited their school closures to 2020, Germany continued to use this anti-pandemic instrument in 2021. Even partial school closures with limited capacity, for example, alternating classes, were in place for much longer in Germany than across the OECD, with only Latvia having defaults higher than Germany in 2021 [12]. This was all the more problematic because no binding endpoint was issued by the political decision-makers for this kind of imposed solidarity [see 13, pp. 27–31].

The social distancing measures had a direct (negative) impact, especially on socio-economically disadvantaged children, who are particularly dependent on the provision of reliable structures and public services (cp. the essay by Weyers and Rigó in this volume). Within the group of all young people who are often sweepingly regarded as the "losers of the pandemic", this group the "losers of the losers" [4, p. 74] so to speak – was hit particularly hard by the restrictions: these restrictions included, e.g. the closure of food banks (*Tafel*) for weeks on end, whose services were regularly needed by approximately 500,000 children [14], furthermore the suspension of services and assistance provided by youth welfare, the suspension of preventive medical check-ups for children (*U-Untersuchungen*, similar to *Your child's medical record* in UK) and school enrolment examinations, which can serve as indicators of developmental disorders and maltreatment, the suspension of visits by family midwives, and strict visiting rules in clinics, which sometimes separated parents and children for weeks, etc. [15].

Children and young people with special needs and their families were particularly marginalised during the coronavirus pandemic: they were considered to be potentially more physically vulnerable to the virus than other children, which required special protective measures and may have made their schooling a lower priority. For many disabled young people, this meant even longer-term exclusion from a regular school day and from much-needed therapeutic support [5, pp. 74–76].

2.2 *Is Childcare "systemrelevant"?*[2]

The *lockdown* in March 2020 in Germany excluded those professionals whose occupational activities were classified as "critical infrastructure", which led to a sudden reassessment of jobs with regard to their social significance in the pandemic [16]. This reassessment included organisations or facilities whose failure or impairment would result in lasting supply shortages, significant disruptions to public safety or other serious consequences [17]. In addition to employees of the healthcare system, were included, for example, employees of the police, the fire brigade and the companies that provided food, water or electricity. However, against the background of the closure of schools and day-care centres, one was confronted with the difficulty that many mothers or fathers working in so-called "system-relevant" areas depended on their children being cared for in schools and day-care centres. To enable these

[2] This German term, which cannot be translated equivalently into English, was introduced into the German debate in spring 2020 to identify those sectors of society that should be exempt from a lockdown [see 13, p. 9].

working parents to carry out their activities, which were considered to be essential service, their children were given privileged care in childcare facilities. Those who provided the care, i.e., educators and teachers, were in this manner also classified as part of the "system-relevant sector". However, education and age-appropriate socialisation as taught and provided in schools and day-care centres were not considered essential in themselves. Rather, these institutions were reduced to the task of maintaining the functioning of the other areas considered indispensable to society. Childcare, on the other hand, was degraded to "second-order system relevance" [see 13, pp. 20–22]. Access to institutional care for children was thus primarily based on the significance of parental jobs and less on the psychosocial needs of the children themselves. Even in the case of risk to the child's well-being, children were only granted the possibility of being cared for in childcare facilities outside their home in exceptional cases (compare, e.g., the corresponding exemption regulation of the German federal state of North Rhine-Westphalia [18]). Child and adolescent psychiatrist Jörg Fegert published a wake-up call in the German Medical Journal (*Deutsches Ärzteblatt*) in April 2020 entitled "Kinderschutz ist systemrelevant" (Child protection is system-relevant, see [19]), in which he drew attention to the psychosocial side effects, especially with regard to threats to children's well-being (cf. the following section "Increased risk to children's well-being").

3 Visibilities in Media Discourse and in Policy Advice

In the crisis posed by the pandemic, the social need and demand for orientation knowledge and scientific policy advice increased [20]. In the absence of sufficient data, statements, framework papers, fact sheets or policy letters, different scientific institutions attempted to assess the risks of the pandemic and to recommend reasonable and appropriate measures to control the course of the pandemic. In addition to advice from scientific institutions that had previously been little known to the public, such as the Robert Koch Institute, which is the government's central scientific institution in the field of biomedicine in Germany, or the German National Academy of Sciences Leopoldina, the knowledge of individual experts, e.g. from virology, was also in demand and omnipresent in society [21]. In the media discourse, expert opinions, perspectives and interests of affected social groups, scientific policy advice and controversies about political decisions amalgamated. In the following section, we will trace some examples of particularly present patterns of argumentation on the role of children in the pandemic.

3.1 School in Focus of Policy Advice

The suspicion that children, who were initially less at risk of severe infections, could equally – and often unrecognised – transmit the infection was the most important argument for the school closures. Therefore, they were not propagated as reactive school closures, but rather as a preventive measure at an early stage of the pandemic. According to the recommendations of the Robert Koch Institute, school closures should contribute to slowing the spread of infection in the population. In addition, school closures as an anti-epidemic measure were intended to help avoid overloading the health system, which could already be observed in other European countries at that time [22]. In the public debate that followed, the perspective on children was mainly narrowed on their role as schoolchildren. "We young people are only seen as pupils. We should learn and learn and learn", pupils summed up in the nationwide JuCo survey [23, p. 4]. This emphasis can also be traced in policy advice, e.g. in the third "Ad-hoc-Statement" of the German National Academy of Sciences Leopoldina in April 2020, which focused on "the psychological, social, legal, educational and economic aspects of the pandemic" [24] and which was criticised for its lack of subject matter expertise [25]: children are addressed here almost exclusively in their role as school children. Schools, on the other hand, are seen primarily in their function as places of learning achievement and educational success, beyond being a place of increased virus spread [24, pp. 13–14]. In contrast, the impact of schools on children's socialisation, psychosocial well-being and child protection is hardly mentioned. Although the special burden on families, especially patchwork families and single parents, are discussed separately, children and adolescents, as a particular risk group for psychological consequences, remain unconsidered [24, p. 10]. Aspects of social inequality are mainly focused on learning outcomes and educational success [24, p. 13]. Following this line of argumentation, the partial opening of schools is recommended only for classes relevant to graduation. However, for younger classes and day care-centres, which were deemed to be less relevant to education, a prolongation of the safekeeping of children without providing an educational program (*Notbetreuung*) was recommended.

3.2 School in Focus of Reporting

In the fever curve of reporting in the German public press, in which the term *Corona* led unchallenged during the first year of the pandemic, there were clear peaks but also reporting troughs over the course of the year. These corresponded with a variety of factors, such as acute infection control measures, rising incidence levels, publications of new studies, holiday periods, etc. With regard to topics relevant to children and young people, in the same period the term *Schulen* (schools) received the most coverage with the highest frequency of mentions in a quantitative comparison: the struggle over school closures or openings and related practical questions was the major child- and youth-specific topic complex (Fig. 1).

The life that polemical terms can develop in an emotional crisis situation such as the COVID-19 pandemic is exemplified by an analysis of the term *Treiber der Pandemie* or *Infektionstreiber* (in English approx. "infection drivers"), which was closely linked to the issue of open schools. It was originally introduced into the

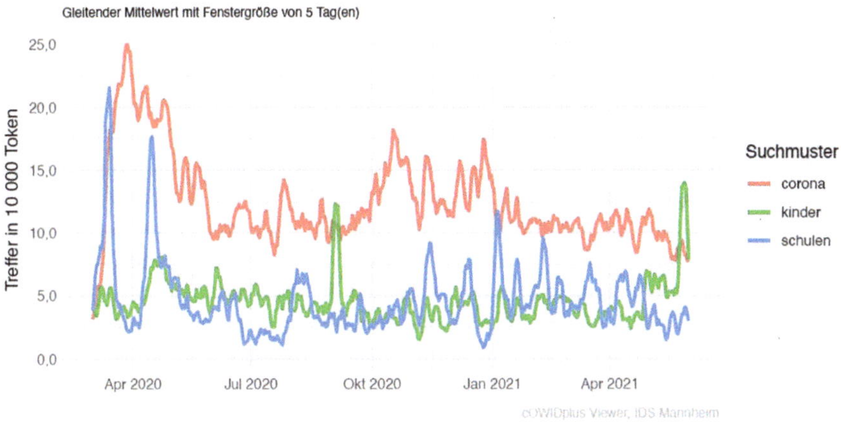

Fig. 1 Fever curves of reporting: Frequency of mention of the terms "Corona", "Kinder" (children) and "Schulen" (schools) in press releases from 13 German-language media. The charts were created with the cOWID plus Viewer of the Leibnitz Institute for the German Language, which can be used to explore the changing vocabulary of German-language (online) press releases in the first year of the Corona pandemic, cp. https://www.owid.de/plus/cowidplusviewer2020. Accessed on August 25, 2023

German debate in June 2020 at a political press conference with the intention of presenting the opening of schools and day-care centres as a scientifically sound, and by no means purely popular, political decision. In view of the renewed increase in the number of infections in autumn 2020, this defensive gesture was questioned and problematised in the media. This was followed by a dispute over the authority to interpret whether these institutions – and the children and young people who attend them – should or should not be regarded as precisely such drivers of infection. The label of *Treiber der Pandemie* was also applied to other areas of social life (e.g., clubs, family celebrations, travellers returning home, restaurants) but initially remained closely linked to the area of schools in 2020. The result of this word-creating, political media joint production was a discursive potpourri of self-exculpation and blaming of others. Contrary to the initial intentions, the stigma of "driver of the pandemic" remained linked to the areas of school and kindergarten for a long time [27]. This narrative temporarily overshadowed the view of school as a central place for children's social development and hindered an objective debate about the necessity and dangers of longer-term school closures.[3]

[3] For a brief summary of the functions and roles, as well as accomplishments and problems, of schools in the Corona pandemic see [4, pp. 81–90].

3.3 Scandalised "Corona Parties"

By way of example, another media phenomenon of the first coronavirus year will be briefly highlighted here, which for a short time gave the group of youths and young adults, who otherwise received little media and political attention, a high – albeit dubious – media presence. In the early summer of 2020, numerous press reports about deliberate violations of the Infection Protection Act (*Infektionsschutzgesetz (IfSG)* [28]) appeared almost simultaneously, orchestrated by warnings and threats from various politicians. The media coverage focused on so-called "coronavirus parties" where young people gathered to celebrate. The purely negative presence of young people in print and on TV in the meantime painted the picture of an irresponsible, egoistic generation which, without any sense of solidarity, was primarily concerned with a hedonistic lifestyle.

The wave of media outrage surrounding the alleged deviant youth behaviour broke as quickly as it had arisen; that the loudly lamented events were by no means systematic in character now seems undisputed [5, pp. 58–65, 29].

There are various patterns of interpretation for the origin and meaning of these accusations, which were exposed as exaggerated soon after they appeared. On the one hand, the criticism of youth could be read as a kind of anthropological reflex which has been passed on from generation to generation. A discursive attempt at restitution of the moral generational order, which had been shaken by the accusations of Fridays-for-future against the "boomer" generation of 2019, also seems conceivable.[4]

Another approach interprets the disproportionate media scandalisation of youth deviance in the early phase of the pandemic as a kind of postmodern 'moral panic'. This phenomenon, described in the 1970s by the sociologist Stanley Cohen (1922–2020), shows clear parallels to the coronavirus parties of 2020. In his influential book "Folk Devils and Moral Panics", Cohen traces the conflict between two rival groups of deviant youths in 1960s England, which was virtually celebrated in the media, scandalised to the point of the grotesque, and followed with an unbelievable wealth of reports – not unlike the reception of the so-called *Halbstarke* in Germany. Cohen interprets the loud stigmatisation of a group outside the bourgeois cosmos as a kind of catalyst for social fears of change. Under the high moral pressure of media outrage, the undesirable, deviant behaviour is massively stigmatised, which in turn leads to increased social control [31].

With regard to the COVID-19 pandemic, the excessive media stigmatisation of a presumed transgression of the contact restrictions in force made clear to the (adult) public the seriousness of these new rules. Following them and sanctioning non-compliance reassured oneself of one's own ethically and morally impeccable status – in contrast to the depraved young subjects who placed themselves outside the bourgeois sphere through their behaviour. For the young people themselves, this moralising alienation of their generation by their elders occupied the discursive

---

[4] For a brief account of the ensuing discourses around intergenerational justice, see [30].

space in relation to their own concerns and needs. In short, those who place themselves outside the community through their "unsolidaric behaviour" need not expect compassion themselves.

4 Making Vulnerability Visible

4.1 Studies on Youth Health and Learning Progress

When was the suspicion substantiated that the closure of educational, care and sports facilities as well as the demands of social distancing would be accompanied by relevant health sequelae in children and adolescents?

As the acute threat posed by the virus subsided through vaccination and new, milder variants and new studies on child health appeared, the initial broad social support for measures such as school closures in the German population[5] from 2022 onwards was slowly eclipsed. It was gradually replaced by the realisation of the physical and, above all, psychological damage that children and adolescents had suffered because of the far-reaching and long-term infection prevention measures. An important role for this change in awareness was due to the results of scientific research on physical and mental health but also by the survey of the learning progress of young people in the (post) coronavirus period:

For example, representative COPSY longitudinal studies have been regularly surveying changes in the mental health status of children and adolescents since the beginning of the pandemic. For this purpose, a comprehensive online survey on the mental health of children and adolescents and their families in Germany was conducted from May to June 2020. Follow-up surveys took place in winter 2020/21, autumn 2021, February 2022 and autumn 2022. The studies showed a decrease in quality of life and a significant increase in mental distress, such as anxiety, depression and sleep problems in children and adolescents over the course of the pandemic, with those from socially disadvantaged families being most affected. After the end of the interventions, these levels decreased but have not yet reached pre-pandemic levels[6]. In a review of 39 current studies on the mental health of children and adolescents during the pandemic, the Robert Koch Institute also found strong pandemic-related stress, increases in mental disorders and impairments in quality of life [33].

Children and adolescents showed greater vulnerability than adults, not only in terms of psychosocial health. The long periods of (partially) closed schools also had a direct impact on their learning progress: a large-scale meta-analysis of 42 studies from 15 countries (mainly studies from the UK and the US, and four studies from Germany) found that learning progress slowed considerably during the coronavirus

[5] The population's approval ratings for relevant pandemic measures such as school closures were evaluated in spring 2020 by a Germany-wide random sample within the "Mannheim Corona Study", see [32].

[6] For the evaluation of the most recent survey, see [3].

pandemic – students lost a total of 35% of the learning progress of a normal school year (the lower the mean average income of a country studied, the worse the results. Since no countries with very low average incomes were included in the study, an even more precarious learning loss must be assumed globally). Children with low socioeconomic status showed the greatest learning deficits, and it is to be feared that many of them could lose touch with our knowledge society with lifelong negative consequences [34]. In the winter of 2022/23, numerous media reports appeared in which representatives from politics and science expressed regret about the negative effects of the infection control measures on children and adolescents [35–39]. The question of whether the central criterion for the German school closures in 2021 was indeed in the best interest of the child is now also discussed by the European Court of Human Rights (ECHR) [40, 41].

4.2 Increased Risk to Children's Well-Being

The extent to which the pandemic has led to more violence in families is still unclear. However, there is growing evidence that the dark field of family violence increased during this time. In May 2020, the first results of online surveys of children and parents indicated that the prolonged closure of childcare facilities was contributing to increased family conflicts in the home environment [42], which was especially true for children in socioeconomically disadvantaged families [43]. Increased parental stress and home quarantine of parents are known to be risk factors for domestic violence [44, 19]. However, youth welfare offices did not receive more reports of alleged child maltreatment – on the contrary, some of them received fewer reports on suspected child maltreatment in the first months of the lockdown [45, 46, pp. 111–112]. In the same period, children's hospitals and child protection outpatient clinics also counted fewer child protection cases compared to the previous year [47]. The decrease in the number of reports on suspected child abuse and neglect, which is an international phenomenon [48, 49], was mainly attributed to the fact that the closures of childcare facilities led to a disruption in standard reporting channels: in Germany, approximately 40% of reports on suspected child maltreatment originated from childcare facilities in the past [50]. The impact of social distancing measures on child protection concerns was underestimated during the first months of the pandemic: therefore, children threatened or even affected by domestic violence, who had previously been partially identified by teachers or educators, became invisible during the time when childcare facilities were closed [51]. The already large dark field of intrafamily violence temporarily developed into a complete black box during the crisis due to the obstructed channels of help, whose immediate – and sometimes deadly – threat to those affected received little resonance either in the media or politically.

As an example, we will refer to the case of 5-year-old Fabio, who was abused to death by his guardians during the first lockdown. The circumstances and antecedents of the crime contain elements that have regularly led to the scandalisation of similar cases in the German media in the past since the "Kevin case" in 2006. In April 2020, media coverage of Fabio's death was limited to a few, mainly regional reports, and there was largely no critical questioning of social and political responsibility [26]. Apparently, the increased reporting on the virus overlapped with other topics in the long term.

4.3 Children's Participation and Rights

Decisions made by politicians during the "gigantic real experiment" ("gigantisches Realexperiment", Michael Hagner [52]) of the coronavirus pandemic were fraught with great uncertainty and considerable risks. Among the particular difficulties was the significant time pressure under which political decisions had to be made. Scientific policy advice also faced similar difficulties, as expressed, for example, in the first Ad-hoc-Statement of the German National Academy of Sciences Leopoldina: "This highly dynamic and unprecedented situation causes uncertainty and requires unconventional solutions, whose impact and unintentional side effects cannot be fully anticipated" [53]. In retrospect, the question arises: Are the currently observed psychosocial burdens of children and adolescents, which are now believed to be a consequence of the various demands of social distancing during the coronavirus pandemic [54] to be regarded as such "unintentional side effects" or could they have been foreseen if the voices of children and their advocates had been perceived during the pandemic?

The media omnipresence of expert opinions on pandemic containment has also generated renewed interest in how scientific policy advice works [55]. In this context, particular attention was drawn to the limits of scientific policy advice and the risk of premature conclusions [56]. Furthermore, the disciplinary membership of policy advisory bodies, their gender representation and the transparency of the process were questioned. Initial studies indicate that women were underrepresented in these bodies. At the same time, the interdisciplinarity of the experts involved was narrowed with a strong dominance of biomedical expertise [57]. However, a lack of representativeness and limited disciplinary diversity in policy consultation jeopardises a fair representation of the interests of all social groups.

Only on the basis of future studies will it be possible to conclusively assess whether the interests of children and young people were sufficiently represented, especially with regard to a risk–benefit assessment of the pandemic measures. Young people themselves expressed that their own social needs were not sufficiently heard or taken notice of during the pandemic, which was reflected in a survey of over 5000 young people conducted by Andresen et al. [6, pp. 14–20]. This finding, which worsened over the course of the pandemic, is not fundamentally new, as previous youth surveys have already shown that young people have felt poorly involved in socially relevant issues and political decisions [6, p. 12]. Opportunities for social and political participation, especially for socioeconomically disadvantaged young people, appear to have been further reduced during the pandemic. Child advocacy groups also criticised the lack of children's voices in the public debate[7]: The German Academy for Paediatrics and Adolescent Medicine as the umbrella organisation of the German paediatric societies,[8] e.g. pointed out that all regulations adopted for children and young people were primarily conceived from the perspective of adults. It was emphasised that experts for children and adolescents were not represented in the political advisory bodies.[9] In spring 2021, criticism about the missing participation of children and young people during the pandemic became part of the debate on anchoring children's rights in the German Basic Law, which was already stated in the coalition agreement in March 2018 [59]. Since the adoption of the UN Convention on the Rights of the Child in 1989, this option has been discussed unsuccessfully on several occasions [60]. However, the treatment of children during the coronavirus pandemic provided new arguments at a time when children were largely banned from the public sphere: "The political discussions and measures taken in Germany have shown that children are quickly overlooked as bearers of independent rights",[10] summarised the Monitoring Body UN Convention on the Rights of the Child in its statement in May 2020. Nevertheless, this legislative initiative also failed due to the lack of a majority in parliament. However, for a brief moment, it did give children's participation rights a political audience.

5 Conclusion

Daniel Kehlmann's opening quotation [1] continues to arouse protest to this day; however, it was prophetic in its core content: at a very early stage of the pandemic, he already saw children at risk, although their own vulnerability was not socially recognised. However, the previously described weighing of interests at the expense of children's health was less an expression of a conscious social decision, as Kehlmann's quotation suggests. Rather, the vulnerability of children was overlooked in favour of a dominant biomedical risk assessment. Furthermore, the narrowed discourse on children whereby children and young people were considered only in specific social roles obscured their own vulnerability. Their psychosocial needs beyond immediate exposure to the virus (e.g. physical proximity, contact with peers, stable care relationships etc.) dropped off the radar, especially with regard to potential side effects of the social-distancing measures, whose duration was initially unknown.

---

[7] Cp. a chronological compilation of statements with regard to child protection during the pandemic: see [58].

[8] This umbrella organisation was recently renamed: "Bündnis für Kinder- und Jugendmedizin". Cp. https://www.buendnis-kjg.de/neuer-name/ (Access on March 18, 2023).

[9] This was noted, e.g. in: Stellungnahme der Deutschen Akademie für Kinder- und Jugendmedizin e.V. zu weiteren Einschränkungen der Lebensbedingungen von Kindern und Jugendlichen in der Pandemie mit dem neuen Coronavirus (SARS-CoV-2), Published online 13 April 2020.

[10] Original quote: "Die politischen Diskussionen und getroffenen Maßnahmen in Deutschland haben gezeigt, dass Kinder als Träger_innen eigenständiger Rechte schnell übersehen werden." [61]

Furthermore, the pandemic discourse indicates that the tasks, functions and services of schools and care institutions are socially underestimated. Their "systemic relevance" – as David Kaldewey described it [see 13, p. 30] for child socialisation, care, protection and education in a broader understanding beyond the acquisition of knowledge became more obvious during the pandemic. This also includes their role in addressing socioeconomic inequalities. Finally, the pandemic revealed not only children's and young people's low chances of participation and involvement in German society, but also their adequate representation by advocates in political bodies has been questioned. The educationalist Johannes Drerup puts it in a nutshell: "The image of childhood that has been conveyed in the public debate and that has implicitly underpinned the Corona policy is that of largely passive addressees, incompetent and insufficiently rational actors compared to adults, who have to comply with state directives and are incapable of participating in democratic decision-making processes".[11] Paradoxically, the anchoring of children's rights in the Basic Law was debated at a time when children's and young people's rights to freedom were severely restricted due to the pandemic. For a future reliable representation of the interests of the youngest members of society, an unbiased and comprehensive view of their psychosocial needs is required.

Core Messages

- From today's perspective, many psychosocial burdens of children are considered to be side effects of the various demands of social distancing during the Corona pandemic.
- The narrowed discourse which considered children and young people only in specific social roles obscured their own vulnerability.
- For a future reliable representation of the interests of the youngest members of society, an unbiased and comprehensive view on their psychosocial needs is required.

References

1. Kehlmann D (2020) Interview: Bauer P, "Es gibt in Zeiten der Angst eine große Bereitschaft zum Gehorsam". Süddeutsche Zeitung Magazin/SZ online, Published online 5 May 2020
2. Arnold N (2020) Wissenschaftliche Politikberatung zur gesundheitlichen Bewältigung der Corona-Pandemie. Analysen & Argumente 394:1–5
3. Ravens-Sieberer U, Devine J, Napp AK, Kaman A, Saftig L, Gilbert M, Reiss F, Löffler C, Simon A, Hurrelmann K, Walper S, Schlack R, Hölling H, Wieler LH, Erhart M (2022) Three

[11] Original quote: "Das Bild von Kindheit, das in der öffentlichen Debatte vermittelt wurde und das der Corona-Politik implizit zugrunde lag, ist das von weitgehend passiven Adressaten, im Vergleich zu Erwachsenen inkompetenten und nicht hinreichend rationalen Akteuren, die den staatlichen Vorgaben Folge zu leisten haben und nicht fähig sind, sich an demokratischen Entscheidungsprozessen zu beteiligen." [4, p. 48].

years into the pandemic: results of the longitudinal German COPSY study on youth mental health and health-related quality of life. Published online 11 January 2023
4. Drerup J (2022) Kinder, Corona und die Folgen: Eine kritische Bestandsaufnahme. Campus, Frankfurt, New York
5. Gravelmann R (2022) Jugend in der Krise. Die Pandemie und ihre Auswirkungen. Beltz Juventa, Weinheim
6. Andresen S, Heyer L, Lips A, Rusack T, Schröer W, Thomas S, Wilmes J (2021) Das Leben von jungen Menschen in der Pandemie. Erfahrungen, Sorgen, Bedarfe. Bertelsmann Stiftung, Gütersloh
7. Wiesemann C (2019) Verletzbarkeit. In: Drerup J, Schweiger G (eds) Handbuch Philosophie der Kindheit. Metzler, Stuttgart, pp 185–190
8. World Medical Association (2013) Declaration of Helsinki – ethical principles for medical research involving human subjects. In: 64th WMA general assembly, Fortaleza, Brazil. Published online October 2013, article 19–20
9. Deutsche Akademie für Kinder- und Jugendmedizin (2020) Empfehlung zum Management von Patienten mit Verdacht auf die COVID-19-Erkrankung durch SARS-CoV-2. Published online 21 March 2020
10. Robert Koch-Institut (2020) COVID-19: Jetzt handeln, vorausschauend planen. Strategie-Ergänzungen zu empfohlenen Infektionsschutzmaßnahmen und Zielen (2. Update). Epidemiologisches Bulletin 12:3–6
11. Presse- und Informationsamt der Bundesregierung (BPA): Vereinbarung zwischen Bundesregierung und den Bundesländern: Leitlinien zum Kampf gegen die Corona-Epidemie vom 16.03.2020 (entered into force on 22 March 2020).
12. OECD (2021) Education at a Glance 2021. OECD Indicators, Paris. https://doi.org/10.1787/3197152b-en
13. Kaldewey D (2020) Was bedeutet Systemrelevanz in Zeiten der Pandemie? Berl J Soziol 32:7–33
14. Tafel Deutschland e.V. (2021) Neue Wege. Jahresbericht 2021. Published online
15. Bujard M, von den Driesch E, Kerstin R, Laß I, Thönnissen C, Schumann A, Schneider N (2021) Belastungen von Kindern, Jugendlichen und Eltern in der Corona-Pandemie. Federal Institute for Population Research (BIB). https://doi.org/10.12765/bro-2021-02
16. Lübker M, Zucco A (2020) Was ist wichtig? Die Corona-Pandemie als Impuls zur Neubewertung systemrelevanter Sektoren. WISI Mitteilungen 73:472–484
17. MAGS-NRW, Ministerium für Arbeit, Gesundheit und Soziales (2020) Leitlinie zur Bestimmung des Personals kritischer Infrastrukturen. Published online 15 March 2020
18. Ministerium für Arbeit, Gesundheit und Soziales des Landes Nordrhein-Westfalen (MAGS), Karl-Josef Laumann, Verordnung zum Schutz vor Neuinfizierung mit dem Coronavirus SARS-CoV-2 im Bereich der Betreuungsinfrastruktur (Coronabetreuungsverordnung – CoronaBetrVO), April 16, 2020, § 1, Abs. 3. Published online 16 April 2020.
19. Fegert JM, Clemens V, Berthold O, Kölch M (2020) Kinderschutz ist systemrelevant. Deutsches Ärzteblatt 117:A 703–A 706
20. Knie A, Simon D (2020) Wissenschaft und Politik als Haftungsgemeinschaft. Lehren aus der Corona-Krise. WZB-Mitteilungen (Wissenschaftszentrum Berlin für Sozialforschung) 168:86–88
21. Böcher M, Krott M (2022) Wissenschaftliche Politikberatung in der Corona-Krise. In: Politik zwischen Macht und Ohnmacht. Studien zur Inneren Sicherheit. Springer, Wiesbaden, pp 305–338
22. Haas W, an der Heiden M, Buda S, Rexroth U (2020) Fachliche Stellungnahme zu Schulschließungen als bevölkerungsbezogene antiepidemische Maßnahme, Epidemiologisches Bulletin 12:7–8. Published online 19 March 2020
23. Andresen S, Lips A, Möller R, Rusack T, Thomas S, Schröer W, Wilmes J (2020) Erfahrungen und Perspektiven von jungen Menschen während der Corona-Maßnahmen. Erste Ergebnisse der bundesweiten Studie JuCo. Universitätsverlag Hildesheim, Hildesheim

24. German National Academy of Sciences Leopoldina (2020) 3rd Ad-hoc-Statement, coronavirus pandemic – sustainable ways to overcome the crisis, p 2. Published online 13 April 2020
25. Wiesing U, Becker D, Hahn P, Tümmers H, Blum CD (2021) Wissenschaftliche (Politik-) Beratung in Zeiten von Corona: Die Stellungnahmen der Leopoldina zur Covid-19-Pandemie. Ethik und Gesellschaft 1/2021: Pandemie-Nach-Denken, p 1–23
26. Griemmert M (2022) Von Virenschleudern und Partygängern: Die mediale (Re-)Präsentation von Kindern und Jugendlichen in der Pandemie. In: Kinder und Jugendliche in der Covid-19 Pandemie: Perspektiven aus Praxis und Wissenschaft. Düsseldorf University Press, pp 119–130
27. Adler D (2021) "Infektionstreiber" im Corona-Diskurs: Der diskursive Kampf um die Lasten der Pandemiebekämpfung. DiscourseNet. Collaborative Working Paper Series, note 4. Published online February 2021
28. Gesetz zur Verhütung und Bekämpfung von Infektionskrankheiten beim Menschen (Infektionsschutzgesetz - IfSG) vom 20. Juli 2000 (BGBl. I S. 1045), das zuletzt durch Artikel 8b des Gesetzes vom 20. Dezember 2022 (BGBl. I S. 2793) geändert worden ist (status of March 2023).
29. Okunew N, Theinert N (2020) Die Pandemie als Generationenkonflikt? Von "Corona-Partys" und Moral Panics. Zeitgeschichte-online. Published online April 2020
30. Fangerau H, Fehlemann S, Griemmert M (2022) Krankheit und Kränkung. Überlegungen zu Seuchen und Invektivität in historischer Perspektive. Seuchen als gesellschaftliche Herausforderung in landeshistorischer Perspektive. Geschichte im Westen. Zeitschrift für Landes- und Zeitgeschichte 37:20–22
31. Cohen S (1972) Folk devils and moral panics: the creation of the Mods and Rockers. MacGibbon, London
32. Blom AG (2022) Zum gesellschaftlichen Umgang mit der Corona-Pandemie. Aus Politik und Zeitgeschichte 70:16–22
33. Schlack R, Neuperd L, Junker S, Eicher S, Hölling H, Thom J, Ravens-Sieberer U, Beyer AK (2023) Changes in the mental health in the children's and adolescent population in Germany during the COVID-19 pandemic – results of a rapid review. J Health Monit 8:2–72
34. Betthäuser BA, Bach-Mortensen AM, Engzell P (2023) A systematic review and meta-analysis of the evidence on learning during the COVID-19 pandemic. Nat Hum Behav, 1–11. https://doi.org/10.1038/s41562-022-01506-4
35. Buyx A, Interview: Kühn K (2022) Alena Buyx: "Die Belange junger Menschen wurden nicht genügend gesehen". Interview in Deutschlandfunk. Published online 28 November 2022
36. Laschet A, Laumann KJ, Machowecz M, Montgomery F, Prien K, Priesemann V, Raether E, Ramelow B, Schmidt-Chanasit J, Schünemann M, Schwesig M, Sentker A et al. (2023) Da habe ich mich geirrt: Vom Ministerpräsidenten bis zur Virologin: 20 Menschen, die in der Pandemie eine wichtige Rolle spielten, gestehen ein, wo sie falschlagen - und was sie heute nicht mehr so machen würden. DIE ZEIT 05/2023:10–11
37. Lauterbach K, Interview: Stempel M (2023) Schließung von Schulen und Kitas: Lauterbach räumt Fehler in Corona-Politik ein. ARD Morgenmagazin. Published online 30 January 2023, 5 min
38. Stark-Watzinger B, Interview: Grimm C, Junginger B (2023) Forschungsministerin zu Krebstherapie: "Noch in diesem Jahrzehnt könnte Durchbruch gelingen". Augsburger Allgemeine. 28 January 2023
39. Wieler L, Interview: Hildebrandt T, Scholz A (2023) Es gab nie nur die Alternative: Entweder weniger Tote oder die Schulen offen halten. DIE ZEIT 05/2023:8–9
40. European Court of Human Rights (ECHR) (2023) Application no. 26657/22. https://hudoc.echr.coe.int/eng?i=001-222473. Published 9 January 2023
41. Stibi B (2023) Schulschließungen: EU-Gericht fordert Erklärung vom Bund. Die Welt am Sonntag 06/2023:1
42. Langmeyer A (2020) Kindsein im Ausnahmezustand. DJI Impulse, Das Forschungsmagazin des Deutschen Jugendinstituts 1:42–43
43. Heintz-Martin V, Langmeyer-Tornier A (2020) Economic situation, financial strain and child wellbeing in stepfamilies and single-parent families in Germany. J Fam Econ Issues 34:1–17

44. Steinert J, Ebert C (2020) Gewalt an Frauen und Kindern in Deutschland während COVID-19-bedingten Ausgangsbeschränkungen: Zusammenfassung der Ergebnisse. Hochschule für Politik München, TUM School of Governance. Published online 2020
45. Mairhofer A, Peucker C, Pluto L, Santen E, Seckinger M, Deutsches Jugendinstitut (2020) Kinder- und Jugendhilfe in Zeiten der Corona-Pandemie. DJI-Jugendhilfeb@rometer bei Jugendämtern. Deutsches Jugendinstitut, München. Published online 2020
46. Wulf D (2022) Kinderschutz im Lockdown? Das Childhood-Haus Düsseldorf – trotz Corona ein sicherer Ort! In: Kinder und Jugendliche in der COVID-19-Pandemie. Perspektiven aus Praxis und Wissenschaft. Walter de Gruyter, Berlin, pp 111–118
47. Heimann T, Ewert J, Metzner F, Sigmund F, Jud A, Pawils S (2021) Medizinischer Kinderschutz während des Corona-Lockdowns. Monatsschrift Kinderheilkunde 169:346–352
48. Rapp A, Fall G, Radomsky A, Santarossa S (2021) Child maltreatment during the COVID-19 pandemic. Pediatr Clin North Am 68(5):991–1009
49. Marmor A, Cohen N, Katz C (2023) Child maltreatment during COVID-19: key conclusions and future directions based on a systematic literature review. Trauma Violence Abuse 24(2):760–775
50. Jentsch B, Schnock B (2020) Child welfare in the midst of the coronavirus pandemic – emerging evidence from Germany. Child Abuse Negl 110:104716
51. Driscoll J, Hutchinson A, Lorekt A, Kiss K, Kinnear E (2021) Hearing the voice of the child through the storm of the pandemic: the impact of COVID-19 measures on the detection of and response to child protection concerns. Int J Childrens's Rights 29:400–425
52. Hagner M (2021) Auf die Probe gestellt. Zum Verhältnis von Wissenschaft und Demokratie. Forschung & Lehre 2:100–102
53. German National Academy of Sciences Leopoldina (2020) 1st Ad-hoc-Statement, Coronavirus Pandemic in Germany: Challenges and Options for Intervention. Published online 21 March 2020
54. Ravens-Sieberer U, Kaman A, Erhart M, Devine J, Schlack R, Otto C (2021) Impact of the COVID-19 pandemic on quality of life and mental health in children and adolescents in Germany. Eur Child Adolescent Psychiatry 31(2022):879–889. Published online 25 January 2021
55. Weingart P (2021) Wissenschaftliche Politikberatung in Krisenzeiten. APUZ. Aus Politik und Zeitgeschichte. Published online 15 January 2021
56. Zilla C (2020) Wissenschaftliche Politikberatung in Zeiten der Pandemie. Herausforderungen für die qualitativ-empirische Sozialforschung, Forschungsgruppe Amerika, Arbeitspapier Nr. 2020/01, Dezember 2020. Published online December 2020
57. Sell K, Saringer-Hamiti L, Geffert K, Strahwald B, Stratil JM, Pfadenhauer LM (2021) Politikberatung durch Expert:innenräte in der SARS-CoV-2-Pandemie in Deutschland: Eine Dokumentenanalyse aus Public-Health-Perspektive, Zeitschrift für Evidenz, Fortbildung und Qualität im Gesundheitswesen Wesen (ZEFQ) 165:1–12. Published online 31 August 2021
58. Nationales Zentrum Frühe Hilfen, Stellungnahmen im Zusammenhang mit Corona. https://www.fruehehilfen.de/service/antworten-auf-praxisfragen/corona-zeiten-empfehlungen-und-faq-fuer-fachkraefte-in-den-fruehen-hilfen/stellungnahmen-im-zusammenhang-mit-corona/. Accessed 18 March 2023
59. Oommen-Halbach A (2022) Verankerung der Kinderrechte im Grundgesetz? Über die Geschichte und aktuelle Konjunktur einer Debatte. In: Kinder und Jugendliche in der COVID-19-Pandemie. Perspektiven aus Praxis und Wissenschaft. Walter de Gruyter, Berlin, pp 131–144
60. Wapler F (2017) Kinderrechte ins Grundgesetz? In: Sachverständigenkommission 15. Kinder- und Jugendbericht, Materialien zum 15. Kinder- und Jugendbericht. Zwischen Freiräumen, Familie, Ganztagsschule und virtuellen Welten – Persönlichkeitsentwicklung und Bildungsanspruch im Jugendalter. Deutsches Jugendinstitut, München
61. Monitoring-Stelle UN-Kinderrechtskonvention. Stellungnahme. Kinderrechte in Zeiten der Corona-Pandemie. Kinderrechtsbasierte Maßnahmen stützen und schützen Kinder und Jugendliche in Krisenzeiten, p 1. Published online May 2020

Maria Griemmert I am a historian and work as a research assistant at the Institute for History, Philosophy and Ethics of Medicine, at Heinrich Heine University Düsseldorf. My research focuses on childhood history and currently on the effects of the Corona pandemic on children and adolescents.

Anne Oommen-Halbach I am a paediatrician and work as a research assistant at the Institute for History, Philosophy and Ethics of Medicine, Heinrich Heine University Düsseldorf, Centre for Health and Society. My research focuses on history and ethics of childhood as well as on the history of paediatrics and child psychiatry.

Chapter 9
Social Inequality in Child Health and Development—Before and After the COVID-19 Pandemic

Simone Weyers and Mariann Rigó

In the face of the pandemic, all are equal
(common parlance)

Abstract Vulnerable children, i.e., children growing up in families with low socioeconomic positions have a higher risk of poor health and developmental problems. This is mainly due to differences in material, psychosocial and behavioural pathways embedded in larger societal structures. Since the pandemic has put even more material and psychosocial pressure on vulnerable families and since health-promoting structures (kindergartens, schools) were no longer available, we observe even more problems in health and development in this group. This is shown in the literature and in our own analyses using data from school enrolment examinations. Prevention efforts should be intensified, especially for children from families with low socioeconomic positions, particularly in the areas of obesity and language development.

Keywords Socioeconomic position · School enrolment examinations · Health and development problems · Children · Prevention · Health promotion

---

S. Weyers (✉) · M. Rigó
Institute of Medical Sociology, Centre for Health and Society, Medical Faculty and University Hospital Düsseldorf, Heinrich-Heine-University Düsseldorf, Düsseldorf, Germany
e-mail: Simone.Weyers@uni-duesseldorf.de

M. Rigó
e-mail: Mariann.Rigo@uni-duesseldorf.de

S. Achella and C. Marazia (eds.), *Vulnerabilities*, Integrated Science 18,
https://doi.org/10.1007/978-3-031-39378-5_9

1 Introduction

Vulnerable children, here: children growing up in families with low socioeconomic position [1, 2], have a higher risk of poor health and developmental problems. This phenomenon, called *health inequalities*, is due to differences in material, psychosocial and behavioural risks or resources that are embedded in larger societal structures. Since the pandemic has put even more material and psychosocial pressure on vulnerable families and since health-promoting structures (kindergartens, schools) were no longer available, we hypothesise that vulnerable children are more affected by the pandemic in terms of health and development than their better-off peers. We will show the development of health inequalities in the course of the pandemic, especially by using data from school enrolment examinations.

2 Child Health and Development on Starting School

In general, the state of children's health in Germany is good. Because of improved living conditions and medical care, infectious diseases and disabilities have declined. In the second wave of the federal child and teen health survey (Kinder- und Jugendgesundheits survey; *kiggs survey*) conducted between 2014 and 2017, 96 of 100 parents classified their children's health as good or very good [3]. Nonetheless, the health of children in the early years of life—to which the present chapter refers—is marked by various problem areas, the so-called *new morbidities*. In these cases, the spectrum of health and development is increasingly characterised by disruptions to mental and functional development, as well as by behaviour-dependent physical disorders [4]. This has most notably resulted in behavioural disorders, developmental disorders in the areas of language, cognition and coordination, as well as overweight and obesity [4].

Information on the extent of these new morbidities is provided, among others, by the medical school enrolment examinations that are obligatory for children starting school in all German federal states. These offer the advantage of covering entire cohorts of preschoolers, thus avoiding the selection biases of representative surveys. Using the North Rhine-Westphalian (NRW) school enrolment examinations of 2018, it has been documented that 28.1% of girls and 30.0% of boys did not have an age-related language development [5]. As such, the most frequently observed developmental disorder relates to language [6]. In the Lower Saxony school enrolment examinations, which assess behavioural irregularities on the basis of emotional problems, problems with peers, behavioural problems and hyperactivity [7], 27.7% of the children were affected [8]. On the basis of the Kiggs survey, fewer girls in this age group were affected (13.8%) than boys (22.3%) [9]. Coordination problems are also among the most frequently observed developmental delays in preschoolers [10]. In 2018, irregularities were observed in 7.3% of girls and 11.0% of boys starting school in NRW [11]. Moreover, 6.3% of girls and 5.7% of boys were overweight, while 4.5% of girls and 4.8% of boys were obese [11].

3 Social Inequalities in Child Health and Development on Starting School

Marked social inequalities exist with regard to child health and development. This is amply demonstrated in the literature. The primary example to cite here is the *kiggs survey*. The representative random sample provides evidence of inequalities in child health and development [12], including those mentioned above. Vulnerable children (here: children with low socioeconomic position—SEP) participate less frequently in logopaedic therapy as an indicator of impaired language development than their peers from better socioeconomic backgrounds [13]. Behavioural irregularities are more frequently observed in these children [14], their body coordination, agility and stamina are poorer [15], and they are more frequently overweight or obese [16].

School enrolment examinations are ideal data sources for analysing health inequalities due to their coverage of all social groups, as mentioned above. Within the scope of the *health at school entrance* study (Gesundheit bei Schuleingang) conducted in Düsseldorf [17], we were able to combine social data, obtained by inquiring parents, with medical examination data of the school starters of 2017 and 2018. Using different categories of parental level of education (as an indicator of SEP) and comparing the prevalence of problems in language, behaviour, coordination and obesity, we observed the expected social gradient. Preschoolers of parents with low education (at most completion of minimum schooling with or without vocational training) were more frequently affected by a.m. problems than their peers whose parents have intermediate-level education. In turn, the latter group was more frequently affected compared to children whose parents have a high level of education (at least one parent with a third-level professional or vocational qualification) [18]. This social gradient is illustrated in Table 1.

Table 1 Social gradient of developmental problems in preschoolers

Parental education	Prevalence in problems regarding			
	Language (%)	Behaviour (%)	Coordination (%)	Overweight (%)
High	15	2	17	6
Intermediate	23	4	23	14
Low	31	6	26	19

4 Potential Explanations for Social Inequalities in Childhood Development

Social inequalities in child and adult health are not a typical German phenomenon. They can be observed in all European countries with different political, economic, cultural and welfare systems [19]. This already shows that unequal access to health care is not a sufficient explanation for health inequalities. In fact, it is a complex mix of various proximal and distal factors. Researchers have produced various models for systemising this phenomenon in adulthood [20, 21] and in childhood [22, 23]. At this point, we refer to a current overview summarising the key indicators of the cause of health inequality in childhood [24]. According to this study, the social structures influencing the family's socioeconomic situation are the primary drivers of health inequalities. They have an effect on a family's socioeconomic situation, for example, by way of state benefit payments or political rulings on working hours, working from home and parental leave. The latter, in turn, give rise to three pathways affecting childhood development:

(i) the availability of material means determines whether families can afford the essentials of daily life, such as adequate living space, food and clothing;
(ii) in psychosocial terms, experiencing the loss of status and control can activate stress responses. With young children, these experiences first manifest through the parents. Teenagers then develop their own understanding of social positioning. However, further daily stressors such as financial pressures are also relevant here. On the other hand, social networks help to relieve the health impacts of these stressors by providing support. Conversely, social isolation can be experienced as oppressive;
(iii) on a behavioural level, factors such as diet, movement, smoking or alcohol consumption are relevant. During pregnancy and in early life, parents' behaviour is key and effective. Later, children and teenagers develop their own health-related behaviours.

Material means, psychosocial resources or risks and health-related behaviours are connected with each other in many ways. Some examples of this are the inability of economically disadvantaged families to afford sport amenities or stress-induced excess eating. The three pathways are furthermore influenced by social structures at a local level. These include the housing condition, the quality of kindergarten and schools, the availability of leisure activities, or the provision of medical care, prevention and health promotion.

5 Social-Differential Impact of the COVID-19 Pandemic on Drivers of Health Inequalities

The following paragraphs present preliminary assumptions and, as far as available, evidence on how the pandemic affected these pathways. We apply a social-differential perspective with regard to the children's material situation, their psychosocial pressures, and their health behaviours.

5.1 Financial Pressures

Even before the COVID-19 pandemic, one in seven children in Germany was threatened by poverty and social exclusion [25], and it may be assumed that the pandemic has further exacerbated this situation. The *Distribution Report* 2020 (Verteilungsbericht) analysing the income situation in connection with the pandemic [26] found that families with lower income had to accept drops in income more frequently and that these losses also had a greater impact. With restrictions on institutional childcare, particularly at the start of the crisis, it was primarily poorly qualified and thus lower-earning employees who did not have the opportunity to work from home. They had to reduce their working hours so that they could take care of their children (ibid). Among the different types of families, single parents were under particular pressure. They face in general a higher risk of poverty: in 2019, 42.7% of all single-parent households were at risk of poverty [27]. Furthermore, they are highly earnings-oriented, and they work more frequently and longer hours than mothers in two-parent families. This earnings orientation is, however, highly dependent on a reliable childcare infrastructure. When kindergartens and schools were closed in March 2020, the problem of childcare was pushed to the private sector, and there were also restrictions on interpersonal contact. While two-parent families were able to split this burden, single parents faced the situation of having to manage work, childcare and daily life without support [28]. As yet, there are no reliable data on the development of the proportion of single parents at risk of poverty after the pandemic, but it may be assumed that the reduction of working hours has had a negative effect on the already tense economic situation of single-parent families. At the same time, research has shown that families in economic difficulties are not receiving state benefits even though they are entitled to them. During the pandemic year 2020, only 55% of some two million people under the age of 15 entitled to benefits received funds from the state's education and inclusion package [29]. Experts have criticised the bureaucratic hurdles involved and are calling on the authorities to actively provide information on statutory benefits [29].

5.2 Psychosocial Stressors

During the pandemic, children experienced a wide range of different stressors. First, the abovementioned financial crises within families have arguably put pressure on them. There is ample proof of this from the first lockdown. Parent surveys show that over one-third of single parents of children aged up to six said that their income as a family was not sufficient during the pandemic [30]. Apart from experiencing status crises or the loss of status, there were also family conflicts at home arising out of contact restrictions. The changes to rules, structures and everyday routines unsettled both parents and children and put pressure on family interactions. In parent surveys, financially challenged parents first and foremost reported an increase in arguments and conflicts in the family, as well as an increase in psychosomatic complaints [30]. A further pressure on (young) children may have been discovering that they were not equipped to meet the needs of home schooling. A press release from the Federal Statistical Office, for example, notes that whether families have the necessary computer equipment is highly dependent on their income. In 2019, while almost half (46%) of households with at least one child and a net income of under 2000 euros did not own a tablet, approximately one-fifth (18%) of households with a monthly income of 5000 to 18,000 euros did [31]. However, for home schooling, children needed not only a computer but also other resources such as a printer to print worksheets at home and a quiet environment so that they could learn, while not disturbing other family members also engaged in home schooling or working from home. A survey of household equipment for living and learning shows that only 64.2% of children in families on minimum unemployment benefits (ALG II) have their own room, while only 69.5% have their own desk [32]. Apart from barriers to accessing digital learning at home, there were also indications that teaching personnel had differing expectations during the pandemic. While teachers in privileged locations more resolutely strove to maintain standards, teachers at disadvantaged schools were more oriented to pupils' emotional needs at the expense of educational standards. Nonetheless, social inequalities are being systematically reproduced in the educational system because content and digital competences cannot be learned [33].

5.3 Health-Related Behaviour

The closure of kindergartens and schools affected all children, but the negative impacts on lifestyle were particularly striking for socioeconomically disadvantaged children. This first relates to physical activity. The closure of kindergartens or schools meant that both everyday physical movement (walking to kindergarten/school, running and playing during breaks) and curricular exercise and sport disappeared. Instead, children—and schoolchildren in particular—spent more or less time at home in front of a screen, depending on the given school's home-schooling strategy. In addition, many physical movement and sport amenities in the leisure sector, such as football, gymnastics or swimming, were suspended for a lengthy period, not to mention the complete closure of kickaround areas for months on end. Although all children were affected, those with less access to alternative play and movement spaces, such as their own garden, parks or playgrounds, were particularly hard hit. Poorer families more frequently live in homes without gardens. In the *Socioeconomic Panel* (SOEP), for example, only 34% of children in families on minimum unemployment benefits said that they had their own garden or had access to one [32]. Studies on the physical activity of children during the COVID-19 pandemic confirm these social inequalities. Houses (versus apartments) were correlated with increased outdoor activities [34], while children from lower-socioeconomic-status families were reported to play outdoors less frequently [35, 36].

The closure of kindergartens and schools also affected eating behaviours. Social inequalities with regard to the consumption of healthy foods [37] and the daily consumption of breakfast [38] in children have been demonstrated before the pandemic. With their health-promoting structures, such as group breakfast at kindergartens or school classes, educational institutions offer socioeconomically disadvantaged children important access to healthy eating. During lockdown, these children lost access to setting-based nutrition programmes [39] and were restricted to their home environments with increased food intake and unhealthy food choices [40].

6 Social-Differential Impact of the COVID-19 Pandemic on Child Health and Development

Overall, the pandemic has influenced children's health and development in many ways. Individual studies and reviews meanwhile point to worsening of language development [41, 42], mental health [43], motor skills and overall physical fitness [44], and weight [39, 40]. However, studies investigating the impact of the COVID-19 pandemic on child health have seldom applied an inequalities perspective—with some exceptions. The few available studies are based on different data sources (e.g., school enrolment examinations, hospital data or representative surveys) and show that children with low SEP have a higher increase in language problems [41], mental health impairments [45, 46], coordination problems [47] and overweight [41, 48] compared to their better off reference group.

Given this lack of evidence, we have used school enrolment examinations of the city of Düsseldorf to systematically analyse the change in child health and developmental (HAD) problems before and after the pandemic [49]. In our trend study, we included four cohorts of preschoolers (school entrance 2018–2021). We computed the predicted prevalences of overweight, coordination and language problems as indicators of HAD problems and analysed their trends separately for vulnerable and nonvulnerable children. Here, vulnerability was operationalised by neighbourhood deprivation as a proxy of SEP. It should be noted that due to data protection, many public health authorities refrain from assessing individual-level social data. In the absence of these, we can make use of a measure describing the sociospatial deprivation of a child's neighbourhood. Sociospatial category systems have been developed in many communities to distinguish and prioritise neighbourhoods in community planning processes, e.g., in child and youth welfare [50]. Based on social data of a given neighbourhood (e.g., number of inhabitants, nationality, living space per inhabitant, number of single-parent households, number of inhabitants on benefits, a category of burden is assigned. Usually, four to five categories are distinguished, ranging from high to low sociospatial burden. Our outcome variables, coordination and language problems, were measured by medical officers in a standardised way [51]. For the analyses, they were dichotomised based on the number of points achieved during the examinations. We used cut-off values indicated by the medical officer [42]. Overweight was operationalised as gender-specific BMI > 90th percentile according to Kromeyer-Hauschild et al. [52]. We computed the predicted prevalences of HAD problems separately for vulnerable and nonvulnerable children using logistic regressions (adjusted for family status, nationality and sex). It should be noted that the 2020 and 2021 samples are not representative of all preschoolers. Vulnerable children (with health problems or from disadvantaged neighbourhoods) are examined with priority each year. Therefore, our survey samples in the pandemic years (due to the limited number of examinations) are smaller and include a selection bias towards vulnerable children. To compensate for that bias, we chose the first 800 children of each cohort, following Bredahl's approach [42].

Figure 1 shows the adjusted prevalences of overweight, language problems and coordination problems in each cohort of preschoolers between 2018 and 2021. We observed a deteriorating trend of HAD problems for all children in the course of the pandemic. Contrary to our expectations, children from well-off districts experienced a larger change compared to those from deprived neighbourhoods. However, among vulnerable children, we found significantly larger prevalences of overweight and language problems in the course of the pandemic in absolute terms. This is not the case in terms of coordination problems, where we observed higher prevalences among better-off children in the last waves of our study. Our analysis [49] also found differing trends of HAD problems among children with non-German versus German nationality and among children growing up in single-parent versus two-parent families. The trends of language developmental problems by nationality background deserve distinctive attention among our results. While both non-German and German children experienced an increase in the prevalence of language problems, non-German children had a marked disadvantage of 30–40 percentage points higher prevalence in each wave.

7 Conclusions

To summarise, vulnerable children had a higher risk of poor health and developmental problems before the pandemic. We have also shown that they suffered more from the consequences of the COVID-19 pandemic than their better-off peers:

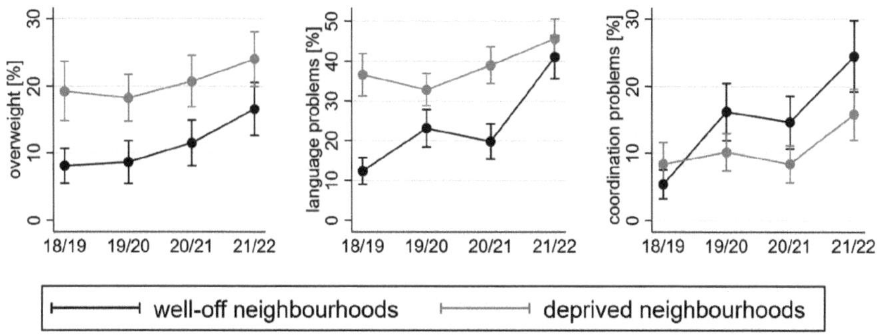

Fig. 1 Predicted prevalences of overweight, coordination and language problems by neighbourhood deprivation and school year

Their families have more frequently experienced loss of income, have been more exposed to psychosocial pressures, and have suffered more from restrictions with a bearing on health behaviours. Consequently, we observed more problems in health and development in these children, partly in the literature and partly in our own analyses. In the latter, against our expectations, the trend is worse in well-off neighbourhoods. However, since the prepandemic prevalences of HAD problems among disadvantaged children were already higher, further deterioration—regardless of magnitude—has led to their particularly poor situation in overweight and language problems.

We conclude that prevention efforts should be intensified, especially for vulnerable children. However, by contrast, there is evidence that in the course of the pandemic, prevention has been challenged in many ways. This is the case for school enrolment examinations [53], parent education and counselling [54, 55], early childhood interventions (*Frühe Hilfen*) [56, 57] and child protection [58, 59]. Although new, digital ways of communication were developed to contact and accompany families, professionals have articulated their concerns. Counselling needs have changed from simple parenting issues towards psychosocial problems such as anxiety or depression, as has been reported by local professionals. Social distancing and lack of personnel, however, have led to decisions of priority. They have shifted the focus towards risk assessment of child endangerment at the expense of the support of many families that are below the risk threshold [59]. It is also stressed by many professionals that digital tools cannot replace personal contact and trust and that new families can never be recruited by this means [56].

Against this background, the revitalisation of established prevention activities is essential. Here, community prevention networks play an important role. These networks (so-called prevention chains, *Präventionsketten* [60], in Germany) aim to promote healthy growth in children, adolescents and young families. Methodically, it is important to avoid top-down approaches and let recipients participate in the prioritisation and development of prevention activities as much as possible. Families with low SEP use prevention services less frequently [61, 62] despite their higher intervention need. This phenomenon is discussed under the term *prevention dilemma* [63]. It is assumed that the distance of disadvantaged groups to prevention services is reinforced if their everyday practices and attitudes are neglected [64]. Conversely, prevention use of disadvantaged groups could be increased by addressing their needs to a larger extent. The approach of community-based participatory research [65] involves both researchers and community members planning, implementing, or evaluating interventions, e.g., regarding obesity [66], physical activity [67], and nutrition [68]. The strength of such an approach lies in the combination of scientific methodological expertise with the practical knowledge of community members [69]. This approach can help mitigate or even remedy the vulnerability of disadvantaged children and their families.

Core Messages

- Vulnerable children have a higher risk of poor health and developmental problems than their better-off peers.
- The pandemic affected and, in some cases, reinforced material, psychosocial and behavioural pathways leading to health inequalities.
- Children in all social groups experienced a deterioration in their health and development over the course of the pandemic. However, there is evidence that vulnerable children ended up with especially high prevalences in overweight and language development compared to their better-off peers.

"Vulnerable children are more affected by the pandemic than their better-off peers."

References

1. Andresen S (2015) Vulnerable Kinder: Interdisziplinäre Annäherungen. Kinder, Kindheiten und Kindheitsforschung Ser, v.10. Springer Fachmedien Wiesbaden GmbH, Wiesbaden
2. Haverkamp F (2018) Gesundheitliche Ungleichheit und neue Morbidität. In: Huster E-U, Boeckh J, Mogge-Grotjahn H (eds) Handbuch Armut und soziale Ausgrenzung, 3. Aufl. 2018. Springer VS, Wiesbaden, pp 479–502
3. Poethko-Müller C, Kuntz B, Lampert T et al (2018) Die allgemeine Gesundheit von Kindern und Jugendlichen in Deutschland: Querschnittergebnisse aus KiGGS Welle 2 und Trends. J Health Monit, 8–15. https://doi.org/10.17886/RKI-GBE-2018-004
4. Reinhardt D, Petermann F (2010) Neue Morbiditäten in der Pädiatrie. Monatsschr Kinderheilkd 158:14. https://doi.org/10.1007/s00112-009-2113-8
5. Landeszentrum Gesundheit Nordrhein-Westfalen (2022) Monitoring Kindergesundheit Reports der Schuluntersuchungen Report 2019
6. Annuß R, Borrmann B, Kuttner S et al, Landesgesundheitsbericht 2015: Informationen zur Entwicklung von Gesundheit und Krankheit in Nordrhein-Westfalen. Ministerium für Gesundheit, Emanzipation, Pflege und Alter des Landes Nordrhein-Westfalen: Referat "Politische Planung, Reden", Düsseldorf
7. Woerner W, Becker A, Friedrich C et al (2002) Normierung und Evaluation der deutschen Elternversion des Strengths and Difficulties Questionnaire (SDQ): Ergebnisse einer repräsentativen Felderhebung. Z Kinder Jugendpsychiatr Psychother 30:105–112. https://doi.org/10.1024//1422-4917.30.2.105

8. Bruns-Philipps E, Hespe-Jungesblut K, Jahn N et al (2018) Kindergesundheit im Einschulungsalter: Ergebnisse der Schuleingangsuntersuchung 2017. Gesundheitsberichterstattung für Niedersachsen
9. Klipker K, Baumgarten F, Göbel K et al (2018) Psychische Auffälligkeiten bei Kindern und Jugendlichen in Deutschland: Querschnittergebnisse aus KiGGS Welle 2 und Trends. J Health Monit 3:37–45. https://doi.org/10.17886/RKI-GBE-2018-077
10. Augste C, Jaitner D, Storr U (2012) Schuleingangsuntersuchung offenbart soziale Unterschiede bei Körperkomposition, Bewegungsverhalten und motorischem Entwicklungsstand. Dtsch Z Sportmed 2012:305–310. https://doi.org/10.5960/dzsm.2012.027
11. Landeszentrum Gesundheit Nordrhein-Westfalen (2021) Monitoring Kindergesundheit. Reports der Schuluntersuchungen. Report 2018
12. Lampert T, Kuntz B (2015) Soziale Ungleichheit der Gesundheitschancen im Kindes- und Jugendalter – Ergebnisse von KiGGS Welle 1. Gesundheitswesen 77. https://doi.org/10.1055/s-0035-1563256
13. Rommel A, Hintzpeter B, Urbanski D (2018) Inanspruchnahme von Physiotherapie, Logopädie und Ergotherapie bei Kindern und Jugendlichen in Deutschland-Querschnittergebnisse aus KiGGS Welle 2 und Trends. J Health Monit 3(4). https://doi.org/10.17886/RKI-GBE-2018-090
14. Kuntz B, Rattay P, Poethko-Müller C et al (2018) Soziale Unterschiede im Gesundheitszustand von Kindern und Jugendlichen in Deutschland – Querschnittergebnisse aus KiGGS Welle 3. J Health Moni 3(2):19–36. https://doi.org/10.17886/RKI-GBE-2018-076
15. Starker A, Lampert T, Worth A et al (2007) Motorische Leistungsfähigkeit. Ergebnisse des Kinder- und Jugendgesundheitssurveys (KiGGS) (Motor Fitness. The results of the German Health Interview and Examination Survey for Children and Adolescents (KiGGS)). Bundesgesundheitsblatt - Gesundheitsforschung - Gesundheitsschutz 50:775–783. https://doi.org/10.1007/s00103-007-0240-8
16. Schienkiewitz A, Brettschneider A-K, Damerow S et al (2018) Übergewicht und Adipositas im Kindes- und Jugendalter in Deutschland: Querschnittergebnisse aus KiGGS Welle 2 und Trends. J Health Monit 3(1). https://doi.org/10.17886/RKI-GBE-2018-005.2
17. Wahl S, Kreffter K, Frölich S et al (2018) Die Schuleingangsuntersuchung als Türöffner für die gesundheitswissenschaftliche Forschung?: Eine Analyse zur Studienteilnahme "schwer erreichbarer" Bevölkerungsgruppen (Is the school entrance examination a door opener for health sciences research? Analysing study participation of hard-to-reach groups). Bundesgesundheitsblatt Gesundheitsforschung Gesundheitsschutz 61:1236–1241. https://doi.org/10.1007/s00103-018-2808-x
18. Weyers S (2022) Soziale Ungleichheit in der kindlichen Entwicklung – Status Quo und Hypothesen zu den Folgen der COVID-19 Pandemie. In: Oommen-Halbach A, Weyers S, Griemmert M (eds) Kinder und Jugendliche in der Covid-19-Pandemie. Perspektiven aus Praxis und Wissenschaft. De Gruyter, Berlin, pp 69–82
19. Mackenbach JP (2019) Health inequalities: persistence and change in modern welfare states, 1st edn. Oxford University Press, Oxford
20. Dahlgren G, Whitehead M (1991) Policies and strategies to promote social equity in health: background document to WHO—strategy paper for Europe, Stockholm
21. Mielck A (2005) Soziale Ungleichheit und Gesundheit: Einführung in die aktuelle Diskussion, 1. Aufl. Programmbereich Gesundheit. Huber, Bern
22. Dragano N, Lampert T, Siegrist J (2010) Grundlagen der Gesundheitsförderung und Prävention in der Kinder- und Jugendhilfe: Wie baut sich soziale und gesundheitliche Ungleichheit im Lebenslauf aus? In: Sachverständigenkommission 13. Kinder- und Jugendbericht (ed) Mehr Chancen für gesundes Aufwachsen: Gesundheitsförderung und gesundheitsbezogene Prävention in der Kinder- und Jugendhilfe. Verlag Deutsches Jugendinstitut, München, pp 11–50
23. Lampert T, Richter M (2009) Gesundheitliche Ungleichheit bei Kindern und Jugendlichen. In: Hurrelmann K, Richter M (eds): Gesundheitliche Ungleichheit: Grundlagen, Probleme, Perspektiven, 2., aktualisierte Aufl. VS, Verl. für Sozialwissenschaften, Wiesbaden, pp 209–230

24. Pearce A, Dundas R, Whitehead M et al (2019) Pathways to inequalities in child health. Arch Dis Child 104:998–1003. https://doi.org/10.1136/archdischild-2018-314808
25. Statistisches Bundesamt (2020) Tag der Kinderrechte: Jedes siebte Kind in Deutschland von Armut und sozialer Ausgrenzung bedroht. Pressemitteilung Nr. N 076 vom 19. November 2020. https://www.destatis.de/DE/Presse/Pressemitteilungen/2020/11/PD20_N076_634.html. Accessed 8 June 2021
26. Kohlrausch B, Zucco A, Hövermann A (2020) Verteilungsbericht 2020: Die Einkommensungleichheit wird durch die Corona-Krise noch weiter verstärkt. WSI Report
27. Bundeszentrale für politische Bildung (2020) Ausgewählte Armutsgefährdungsquoten | bpb. https://www.bpb.de/nachschlagen/zahlen-und-fakten/soziale-situation-in-deutschland/61785/armutsgefaehrdung. Accessed 5 July 2021
28. Stahl A (2021) Corona-Krise trifft Alleinerziehende besonders hart: Pressemeldung vom 26. February 2021
29. tagesschau (2021) Bildungs- und Teilhabepaket: Hilfe kommt bei vielen Kindern nicht an. https://www.tagesschau.de/investigativ/monitor/benachteiligung-kinder-101.html. Accessed 8 June 2021
30. Schäfermeier E, Bremerich L, Gormanns Y et al (2022) Familienalltag in Zeiten von Corona: Zur Betreuung von Kindern und Stress von Familien. In: Oommen-Halbach A, Weyers S, Griemmert M (eds) Kinder und Jugendliche in der Covid-19-Pandemie. Perspektiven aus Praxis und Wissenschaft. De Gruyter, Berlin, pp 1–16
31. Statistisches Bundesamt (2020) Homeschooling: Digitale Ausstattung in Familien hängt stark vom Einkommen ab. https://www.destatis.de/DE/Presse/Pressemitteilungen/2020/07/PD20_N042_639.html. Accessed 13 July 2021
32. Geis-Thöne W (2020) Häusliches Umfeld in der Krise: Ein Teil der Kinder braucht mehr Unterstützung. Ergebnisse einer Auswertung des Sozio-oekonomischen Panels (SOEP). IW-ReportUR. https://www.econstor.eu/handle/10419/216213
33. Bremm N (2021) Bildungsbenachteiligung in der Corona-Pandemie. Erste Ergebnisse einer multiperspektivischen Fragebogenstudie. PraxisForschungLehrer*innenBildung: PFLB: Zeitschrift für Schul- und Professionsentwicklung 3:54–70. https://doi.org/10.25656/01:22186
34. Mitra R, Moore SA, Gillespie M et al (2020) Healthy movement behaviours in children and youth during the COVID-19 pandemic: Exploring the role of the neighbourhood environment. Health Place 65:102418. https://doi.org/10.1016/j.healthplace.2020.102418
35. Poulain T, Vogel M, Sobek C et al. (2019) Associations between socio-economic status and child health: findings of a large German Cohort Study. Int J Environ Res Public Health 16. https://doi.org/10.3390/ijerph16050677
36. Perez D, Thalken JK, Ughelu NE et al (2021) Nowhere to go: parents' descriptions of children's physical activity during a global pandemic. Front Public Health 9:642932. https://doi.org/10.3389/fpubh.2021.642932
37. Kuntz B, Waldhauer J, Zeiher J et al (2018) Soziale Unterschiede im Gesundheitsverhalten von Kindern und Jugendlichen in Deutschland – Querschnittergebnisse aus KiGGS Welle 2. RKI-Bib1 (Robert Koch-Institut)
38. Kuntz B, Giese L, Varnaccia G et al. (2017) Soziale Determinanten des täglichen Frühstücksverzehrs bei Schülerinnen und Schülern in Deutschland: Ergebnisse aus KiGGS Welle 1. In: Gemeinsame Jahrestagung – der Deutschen Gesellschaft für Epidemiologie e.V. (DGEpi), – der Deutschen Gesellschaft für Medizinische Soziologie e.V. (DGMS) und – der Deutschen Gesellschaft für Sozialmedizin und Prävention e.V. (DGSMP). Georg Thieme Verlag KG
39. Chaabane S, Doraiswamy S, Chaabna K et al (2021) The impact of COVID-19 school closure on child and adolescent health: a rapid systematic review. Children (Basel) 8. https://doi.org/10.3390/children8050415
40. Stavridou A, Kapsali E, Panagouli E et al (2021) Obesity in children and adolescents during COVID-19 pandemic. Children (Basel) 8:135. https://doi.org/10.3390/children8020135
41. Bantel S, Buitkamp M, Wünsch A (2021) Kindergesundheit in der COVID-19-Pandemie: Ergebnisse aus den Schuleingangsuntersuchungen und einer Elternbefragung in der Region Hannover (Child health in the COVID-19 pandemic: results from school entry data and a

parent survey in the Hanover region). Bundesgesundheitsbl 64:1541–1550. https://doi.org/10.1007/s00103-021-03446-2
42. Bredahl R (2022) Corona-Pandemie: Entwicklungsprobleme bei Schulneulingen nachweisbar. In: Oommen-Halbach A, Weyers S, Griemmert M (eds) Kinder und Jugendliche in der Covid-19-Pandemie. Perspektiven aus Praxis und Wissenschaft. De Gruyter, Berlin, pp 83–90
43. Wendt Viola T, Lahorgue Nunes M (2022) Social and environmental effects of the COVID-19 pandemic on children. Jornal de Pediatria, S4–S12
44. Pajek SV, Trajkovic N (2022) Impact of the COVID-19 pandemic on the motor development of schoolchildren in rural and urban environments. Biomed Res Int 2022:8937693. https://doi.org/10.1155/2022/8937693
45. Ravens-Sieberer U, Erhart M, Devine J et al (2022) Child and adolescent mental health during the COVID-19 pandemic: results of the three-wave longitudinal COPSY study. J Adolesc Health. https://doi.org/10.1016/j.jadohealth.2022.06.022
46. Monnier M, Moulin F, Thierry X et al (2021) Children's mental and behavioral health, schooling, and socioeconomic characteristics during school closure in France due to COVID-19: the SAPRIS project. Sci Rep 11:22373. https://doi.org/10.1038/s41598-021-01676-7
47. Wessely S, Ferrari N, Friesen D et al. (2022) Changes in motor performance and BMI of primary school children over time-influence of the COVID-19 confinement and social burden. IJERPH 19. https://doi.org/10.3390/ijerph19084565
48. Jenssen BP, Kelly MK, Powell M et al (2021) COVID-19 and changes in child obesity. Pediatrics 147. https://doi.org/10.1542/peds.2021-050123
49. Weyers S, Rigó M (2023) Child health and development in the course of the COVID-19 pandemic: are there social inequalities? Eur J Pediatr (in press). https://doi.org/10.1007/s00431-022-04799-9
50. Bartling L, Reher A-K (2019) Konstrukt Sozialraum: Kleinräumigkeit nutzen und analysieren: Eine Arbeitshilfe für Kommunen und Träger. Arbeitspapier Armutsbekämpfung und Sozialplanung
51. Petermann F (2009) Sozialpädiatrisches Entwicklungsscreening für Schuleingangsuntersuchungen: SOPESS; theoretische und statistische Grundlagen zur Testkonstruktion, Normierung und Validierung. LIGA. NRW, Düsseldorf
52. Kromeyer-Hauschild K, Moss A, Wabitsch M (2015) Referenzwerte für den Body-Mass-Index für Kinder, Jugendliche und Erwachsene in Deutschland. Adipositas - Ursachen, Folgeerkrankungen, Therapie 09:123–127. https://doi.org/10.1055/s-0037-1618928
53. BVÖGD (2021) Schuluntersuchungen fallen wegen Corona aus –soziale Benachteiligung nimmt zu: Pressemitteilung vom 21. Juni 2021, Berlin
54. Juncke D, Lehmann K, Nicodemus J et al (2021) Familienbildung und Familienberatung in Deutschland: Eine Bestandsaufnahme. Freiburg, Düsseldorf
55. Lüken-Klaßen D, Neumann R, Elsas S (2020) kontakt.los! Bildung und Beratung für Familien während der Corona-Pandemie. Otto-Friedrich-Universität
56. Scharmanski S, van Staa J, Renner I (2020) Aufsuchende Familienbegleitung in der COVID-19-Krise durch Gesundheitsfachkräfte der Frühen Hilfen. Die Hebamme 33:20–28. https://doi.org/10.1055/a-1213-8025
57. Renner I, van Staa J, Neumann A et al (2021) Frühe Hilfen aus der Distanz – Chancen und Herausforderungen bei der Unterstützung psychosozial belasteter Familien in der COVID-19-Pandemie (Early childhood intervention from a distance-opportunities and challenges in supporting psychosocially stressed families in the COVID-19 pandemic). Bundesgesundheitsblatt Gesundheitsforschung Gesundheitsschutz 64:1603–1610. https://doi.org/10.1007/s00103-021-03450-6
58. Mairhofer A, Peucker C, Pluto L et al (2020) Kinder- und Jugendhilfe in Zeiten der Corona-Pandemie: DJI-Jugendhilfeb@rometer bei Jugendämtern
59. Witte S, Kindler H (2022) Kinderschutz in Zeiten von Corona – Informelle Angebote und niederschwellige ambulante Hilfen während der Pandemie. Diskurs Kindheits- und Jugendforschung/Discourse Journal of Childhood and Adolescence Research 17:57–71. https://doi.org/10.3224/diskurs.v17i1.04

60. Richter-Kornweitz A (2016) Werkbuch Präventionskette Gesund aufwachsen für alle Kinder! Herausforderungen und Chancen beim Aufbau von Präventionsketten in Kommunen. Landesvereinigung für Gesundheit & Akademie für Sozialmedizin Niedersachsen e.V., Hannover
61. Faßbender DM, Kreffter K, Götz S et al (2022) Is the level of motor development at school entry related to the use of municipal exercise programs? A social-differential analysis. IJERPH 19. https://doi.org/10.3390/ijerph19053047
62. Eickhorst A, Schreier A, Brand C et al (2016) Inanspruchnahme von Angeboten der Frühen Hilfen und darüber hinaus durch psychosozial belastete Eltern (Knowledge and use of different support programs in the context of early prevention in relation to family-related psychosocial burden). Bundesgesundheitsblatt Gesundheitsforschung Gesundheitsschutz 59:1271–1280. https://doi.org/10.1007/s00103-016-2422-8
63. Bauer U (2005) Das Präventionsdilemma: Potenziale schulischer Kompetenzförderung im Spiegel sozialer Polarisierung, 1st ed. 2005. VS Verlag für Sozialwissenschaften; Imprint: VS Verlag für Sozialwissenschaften, Wiesbaden
64. Franzkowiak P (2020) Prävention und Krankheitsprävention. https://www.leitbegriffe.bzga.de/alphabetisches-verzeichnis/praevention-und-krankheitspraevention/. Accessed 23 Oct 2018
65. Minkler M, Wallerstein N (eds) (2008) Community-based participatory research for health: from process to outcomes, 2nd edn. Wiley Desktop Editions. Jossey-Bass, San Francisco, CA
66. Berge JM, Jin SW, Hanson C et al (2016) Play it forward! A community-based participatory research approach to childhood obesity prevention. Families Syst Health J Collaborative Family Healthcare 34:15–30. https://doi.org/10.1037/fsh0000116
67. DeBate RD, Koby EJ, Looney TE et al (2011) Utility of the physical activity resource assessment for child-centric physical activity intervention planning in two urban neighborhoods. J Commun Health 36:132–140. https://doi.org/10.1007/s10900-010-9290-1
68. Stang J, Bonilla Z (2017) Factors affecting nutrition and physical activity behaviors of Hispanic families with young children: implications for obesity policies and programs. J Nutr Educ Behav. https://doi.org/10.1016/j.jneb.2017.08.005
69. Wolff T, Minkler M, Wolfe SM et al (2016) Collaborating for equity and justice: moving beyond collective impact

Simone Weyers I am medical sociologist at the Heinrich-Heine-University Düsseldorf, Centre for Health and Society. My research focuses on child health and prevention in the context of social inequalities. Using routine data such as the school enrolment examination, I analyse different aspects of health inequalities in children and derive prevention priorities and strategies. There is great need for that, since the pandemic has aggravated existing problems of vulnerable children.

Mariann Rigó I am an economist working at the Heinrich-Heine-University Düsseldorf, Centre for Health and Society. In my research, I analyse empirical data using quantitative methods. In addition to my main research interests in labor and health economics, I have participated in several projects analysing the health consequences of the pandemic.

Chapter 10
The (Crip) Art of Reworking Vulnerability—And Perhaps, to Find a Way Out of It

Chiara Montalti

No one moves without a supportive environment and set of technologies. And when those environments start to fall apart or are emphatically unsupportive, we are left to "fall" in some ways, and our very capacity to exercise most basic rights is imperiled. And we could certainly make a list of how this idea of a body, supported yet acting, supported and acting, is at work implicitly or explicitly in any number of political movements.
Judith Butler, Rethinking Vulnerability and Resistance, 2016

Abstract The aim of this essay is to mobilize the concept of disability and vulnerability, especially when they intertwine. For so doing, I will refer to experiments of care collectives as a fertile case study. The concept of interdependence will accompany the analyses of the essay—and I will account both for the richness and for the limitations it entails. First, I will address the debate on vulnerability, mainly within Feminist Theory and Disability Studies. I will take into account the risks of imposing vulnerability upon specific social groups—for example, disabled people—as it can reinforce their disempowerment and minority position. I will then examine vulnerability as an ontological and contextual phenomenon, highlighting how both are especially productive in critically addressing disability. I will also take into account how disability and vulnerability are frequently entrenched in dependence: in this regard, I will underline how care relationships, which are often considered maximum examples of dependency, can both enhance and reduce the vulnerability that can be experienced by disabled people. I will pinpoint how the narrative of vulnerability can produce neglect of disabled people's knowledge and skills, also in the context of care. Therefore, I will focus on the value of their expertise in this field, examining examples of collective care crafted in recent years and—despite the disabling and threatening nature of the event—even implemented during the COVID-19 pandemic.

C. Montalti (✉)
University of Salento, Lecce, Italy
e-mail: chiara.montalti@unisalento.it

S. Achella and C. Marazia (eds.), *Vulnerabilities*, Integrated Science 18,
https://doi.org/10.1007/978-3-031-39378-5_10

Keywords Disability · Care collectives · Mutual aid · Interdependence · COVID-19 · Relationality · Resourcefulness · Adaptability · Ableism

1 Introduction

The aim of this essay is to mobilize the conceptual and practical intercourses between the phenomenon of disability and vulnerability, addressing collective care relationships as a case study. As I will clarify in the essay, vulnerability and disability are not coextensive concepts: it is possible, and rather frequent beyond human rights discourse, to examine the experience of the latter without addressing the former. In this essay, however, it appears epistemologically and politically useful to intersect them, provided that we are faced with a different mobilization of the two concepts as opposed to the most widespread theoretical analyses. I do not aim to flatten both concepts in a homogeneous similarity, but it is relevant to underline how they share a relational nature and how positive social responses can positively bear on both experiences.

I will present some points within the debate on vulnerability that may be useful throughout my analyses. Vulnerability can concern different systems and entities—the environment, technologies, etc.—and it is clear from the topic of this essay that I will exclusively consider human subjects. In particular, I will address the risks vulnerability may entail when it is used to identify certain social groups or individuals, and I will examine the perspectives that underline its ontological and/or contextual nature—especially within disability studies. However, even though it may seem rather surprising from an external perspective, it is important to note that the topic of vulnerability does not play a central role within disability studies, especially in segments more receptive to critical theory and cultural studies. When it is addressed, its controversial nature is highlighted. I will provide some coordinates of the debate and offer possible intersections between disability and vulnerability, which do not position disabled people in a position of minority and disempowerment. In particular, I describe both phenomena as relational and context-related. In this regard, I will refer to feminist theory as well.

The concept of interdependence will guide the essay, as it adequately embodies the intermingling nature of bodies and subjectivities. In this way, interdependence is connected to disability and vulnerability. It also holds, as it will become clear, a peculiar place within theories and practices implemented by disability studies and disabled activism.

The common perception of disabled people as vulnerable often leads to them being seen as in need of assistance, but this can overlook the fact that disabled individuals can also be skilled and trained in the space of *care*. Therefore, it is important to recognize and value the expertise and contributions that disabled people can bring to the field of caregiving, rather than simply assuming that they are the ones who need care. With this aim in mind, I will conclude the essay by examining examples of collective care, which are also crafted in response to social injustices, that *fragilise* disabled people—especially in moments of crisis, such as during the COVID-19 pandemic. While it is fundamental to consider this event as disabling and life-threatening, especially with inadequate healthcare support, it is also urgent to recognize disabled people's expertise. Not only must they not be framed as *disposable*, but they also represent a resource.

Methodologically, I will explore both disability and vulnerability with a transversal approach, which keeps in mind the cultural, social, anthropological and economic factors and, at the same time, consider the embedded, embodied, and material experiences. Disabled people can possibly experience more pain and challenges due to their impairments or the inadequacy of the environment. They can also experience situated vulnerability, especially when their needs are not met, and the value of their lives is neglected—as often happened during the COVID-19 pandemic. Vulnerability, (dis)ableism, and some negative aspects (such as pain) that may characterize some forms of disability can be minimized by positive social responses and a network of care *crip*[1] kinships, without neglecting the personal and material experience of marginalization and sufferance. Within this framework, I will consider disabled people's practical responses to increased vulnerability.

2 Notes on Vulnerability and Disability

First, vulnerability is generally framed as a disposition that potentially exposes a subject to some form of harm or disadvantage. However, there are expressions of vulnerability that inevitably actualize and can therefore be understood as an inherent characteristic of living organisms: every one of them eventually dies and probably at some moment has passed through phases of compromised health. The harm that the subject may endure is not only located in the realm of life and death: vulnerability can qualify as exposure to tangential or different risks, which are, for example, related to sociality, access to education, emotional well-being, financial status, or sharing of data.

Vulnerability has been addressed, on a theoretical level, mostly by Feminist Theory [1–5] or in discourses on human rights in response to crises and social injustices. As shown by contemporary debate, employing the language of vulnerability with regard to specific subjects or social groups can serve diverse agendas, not just progressive ones as it may initially appear but also "paternalistic, racist, misogynist, homophobic, and anti-feminist ones" [6, p. 5, 7]. Even when the intention is to protect specific pockets of society, the outcomes may be unexpected, ambivalent or explicitly unsuitable. It can draw our attention towards injustice and violence but can produce even more vulnerability. Vulnerability can also entail an othering move: some social groups are distanced from a presumably not (or not so much) vulnerable humanity.

[1] Crip, a contract form of the derogative 'cripple', is a term reclaimed by some members of the disability community. It is rooted in activism and politicization of disability.

In this sense, it risks further stiffening social positionings, rather than "mobilizing" social change.

The semantic network of vulnerability generally includes, in a non-exhaustive list, dependency, care, marginalization, injury, subjugation, trauma, and risk [6, 8]. Some of them refer more strictly to a materialist experience, recalling how the flesh can violently collide with the world (trauma, risk, injury), whereas others refer more closely to structural violence and injustice (marginalization, subjugation). Dependency and care are not necessarily negative experiences, even though interdependence—a concept I will further examine—may more fruitfully represent our enmeshed lives, without the derogatory aspects sometimes attributed to being dependent. The mentioned key concepts co-assemble each other, entangling in diverse ways, orbiting around vulnerability: for example, marginalized people tend to occupy a space of medical vulnerability too (e.g., members of the LGBTQIA+ community can have worse access to healthcare).

In this semantic network, I include disability as well; throughout the essay, I will explore the possible negotiations with this association. The concept of vulnerability can "react" in different ways in regard to contact with disability. People are inherently vulnerable *to* disability: they age, have accidents, and predispositions and circumstances add up in a certain way throughout their lives. On the other hand, disability and vulnerability may represent two overlapping experiences: disabled people are often seen as a vulnerable population, both because of their presumed "frailer" bodyminds and because of their marginalized position in society: they appear more susceptible to harm and social injustices. The link [5, 9, 10] between vulnerability and dependence has been widely explored, and it is not difficult to understand why disabled people are often seen as particularly vulnerable. Emphasis is often placed on how their very survival may depend on a dense network of trans-individual ties [11, 12]. For example, someone generally framed as more vulnerable, such as disabled or elderly people, appears dependent on others. At the same time, however, care relationships themselves can both *create* and *compensate for* vulnerability, and I will examine the latter case in the last section.

Disability scholars that examine vulnerability tend to define it as an ontological and universal characteristic of human beings: as already mentioned, human bodies are inherently frail, as they may fall ill, be in pain, and be exposed to external threats. Feminist theorists, as well, underline this transversal form of vulnerability, pinpointing the contemporary social and cultural tendency—in philosophical and critical thought as well—to remove it. This unravelment concerning human nature works as a starting point to rethink the subject in its relational dimension against "masculinist fantasies of sovereign mastery" [13, p. 3]. The very embedded and embodied nature of the subject produces her/his vulnerability [14]. In both feminist theory and disability studies, this framework sustains a counternarration with respect to Western ideas of autonomous, autarchic, and self-contained subjectivities, which shape a hierarchy that posits disabled bodyminds on the disadvantageous section of the spectrum [8, 15–19]. Therefore, unveiling the ontological condition of vulnerability helps us to challenge ableist assumptions about bodyminds.

Within feminist theory and disability studies, the ontological perspective of vulnerability tends to unfold parallel to the situational perspective: it is also interpreted as a process, actualized or not, that depends on context rather than being linked to specific individuals. These fields tend to consider vulnerability as *also* an ecological phenomenon: there are vulnerable *positionings* of subjects and not vulnerable subjects. The concept can be framed as a relationship "to a field of objects, forces, and passions that impinge on or affect us in some way" [13, p. 19, 20]. It describes, in a potentially negative sense, the impact that the context (social, public, institutional...) *may* have on a subject. Although there is not a complete rejection of the category, it is highlighted how risky it might be to impose this *identity* on disabled people. This approach, which situates vulnerability in the inadequacy of networks of accommodation, support, and care, is exemplified by activist and writer Alice Wong:

> When you are disabled and rely on public services and programs [such as Medicaid], you face vulnerability every day. This vulnerability is felt in my bones and in my relationship with the state... The fragility and weakness of my body, I can handle. The fragility of the safety net is something I fear and worry about constantly. [21][2]

As noted by bioethicist and disability scholar Jackie Leach Scully, vulnerability can be associated with characteristics such as lack of agency, "immaturity, weakness, helplessness, passivity, victimhood, humiliation" [18, p. 210], which risk reinforcing the representations already typically directed towards disabled people. As Judith Butler et al. claim, "there is always something both risky and true in claiming that women or other socially disadvantaged groups are especially vulnerable" [7, p. 15]. Associating disability with vulnerability might be dangerous, as it can reinforce a history of marginalization, partial participation in citizenship, and disempowerment. Disabled people have always been subject to various forms of paternalism—by the state, at a political level, in the social arena, in medicine and healthcare—and therefore they challenge any structure not only that excludes them but also that includes them as passive objects (as, for example, associationism led by nondisabled people). I am not claiming that disabled people are never vulnerable, and should never be considered as such: I only wish to reject that they are vulnerable in a *specific* way and that their vulnerability is supposed to be interpreted in *exceptional* terms. This perspective seems to crystallize the already rigid distinction between disabled and nondisabled people [8, 15, 18, 22].

3 First Keyword: Relationality

Disability studies underline from the very beginning how a purely medical and individual account of disability misses the point. In the so-called medical-individual model, disability embodies an individual tragedy and something that concerns the body or the mind of the subject and, whenever possible, must be cured. Disability

[2] In the USA, Medicaid is the federal and state program that provides health insurance for people and families with low income.

is frequently interpreted as a fixed category: one is disabled, or not; disability is a dysfunction, a cluster of deficits, or a form of noncompliance to a determined standard. *You know it when you see it.* It is rarely presented as dynamic, changing, and subordinating to the context. The social and political-relational models of disability have confronted this predominant perspective, highlighting how it represents a phenomenon that originates in the material encounter of a bodymind in a given context [23–25]. As Alison Kafer underlines, disability is a political concept, open to debate, as opposed to a monolithic truth of bodyminds. It is enmeshed in politics and personal relationships, coconstituted in social and cultural processes, and constantly transformed. This perspective moves it from the mere medical-individual realm—however important—to multiple locations instead.

> [D]isability no longer resides in the minds or bodies of individuals but in built environments and social patterns that exclude or stigmatize particular kinds of bodies, minds, and ways of being. [...] The problem of disability is solved not through medical intervention or surgical normalization but through social change and political transformation. [23, p. 6]

It is therefore fundamental to frame the experience of disability in interrelational connections and in sociocultural and political structures, rather than in the narrow boundaries of individuality. Disability emerges from specific normative contexts within medicine, psychiatry, and healthcare discourses and disability from the sociocultural fabric but is also actively metamorphosed by disabled people. It is "experienced in and through relationships; it does not occur in isolation" [23, p. 8]. In addition, it intersects with other axes (such as gender, class, race), which change a subject's experience in a given society.

Giving disability is *also* a sociocultural and political interpretation; however, it must never produce a removal of individual experiences of loss, pain, difficulty, limitation, and failure. They must remain legitimate and shareable. In addition, disability may still be rejected by the subject and not exclusively because of internalized ableism or structural marginalization [26, 27]. Material experiences of disability, which *can* also include bodily and cognitive sufferance (for example, mental pain; chronic fatigue; the sometimes painful interaction with medical-assistive technologies), must never be overlooked. The fundamental point is that this aspect must be incorporated within a heterogeneous framework, in which it is expected that joy, desire, pleasure, forms of creativity and competence, specific expertise, love for the disabled community, and ways of flourishing occur and are realized as well [23, 28]. Analyses of the sociocultural and economic production of disability should not be raised at the risk of delegitimating complex and painful individual experiences. Medicalization and anthropologization can both be dehumanizing, favouring the analyses of structures and frameworks upon the experience of the subjects [29].

In addition, the same attention must be reserved for collective experiences of disablist and ableist structure and for global injustices that produce disablement. Disabled people may not receive fair accommodations and accessibility responses and may not be guaranteed appropriate healthcare, education, or assistance. At the same time, their condition might originate from unsuitable working and living conditions, lack of adequate healthcare assistance, scenarios of war, and environmental crises.

While vulnerability is hardly seen as something to be proud of, as frequently occurs in relation to disability by the disability community [26, 30], the two phenomena can be addressed by moving from similar theoretical and political assumptions. Firstly, they can both be recalled as sources of knowledge and expertise [6, 7, 31, 32]. The same path is followed in this essay, even though with due caution. This attention to adaptation and resourcefulness should be balanced by attention to possible experiences of harm, pain, and difficulties on a material and social level. As Anu Koivunen et al. note,

> It is indeed worth asking how productive the approach to vulnerability […] as simultaneously involving resistance can be when addressing issues like global racism and massive inequalities in basic resources and the ability to live on. What resistance could there possibly be in the utmost realization of vulnerability – death? [6, p. 8]

Within the "massive inequalities" mentioned, it is important to include disablism and ableism. Secondly, at the same time, it is important to consider both individual experiences and social structure analyses.

Last, as already clarified, vulnerability and disability are profoundly relational: the subject's experience in both regards fluctuates depending on the outer reality. As Rosemarie Garland-Thomson claims about disability, "shifting [it] from an attributed problem in the body to a problem of social justice was theoretically groundbreaking": vulnerability, as well, can be more productively addressed as such [33, p. 592]. The setting can *disable* us and make us vulnerable: specific worlds can be more accessible than others, and the structure of society dramatically influences bodyminds' experiences. The experience of both vulnerability and disability can be shaped by the interdependence that characterizes our living as humans: adequate responses can reduce harm, safeguarding personal well-being and guaranteeing a more just society. These responses can include both personal and communal aspects: a fair distribution of wealth, opportunities, and healthcare but also a platform of shared values, which refuse discrimination and reward care, attention, and support.

4 Second Keyword: Interdependence

As already pointed out, vulnerability does not hold a particularly relevant place within disability studies and disabled activism: this does not mean that the experience of it and the surrounding topics are not addressed, but the conventional literature on it is rarely discussed. A key topic, on the other hand, is represented by the concept and *practice* of interdependence, which shapes the disability community. This practical, theoretical and political response against individualism, precarity, and social injustice is privileged over vulnerability because the latter risks being interpreted as a fixed status of the subject rather than a *movement*. In other words, what people do is more important than what they are supposed to be. The reflection on interdependence—what it implies, how it can be crafted, etc.—is a key contribution of disability studies and disabled activism: as a theoretical analysis but also as a practical skill to be explored and improved within community organizing (for example, through mutual aid networks, collective care, support groups). The term emphasizes the interaction between two or more subjectivities and highlights the flux of empowerment and agency—rather than pushing a hierarchical logic, as 'dependence' might do—or giving prominence to individual capacity and willpower—as 'independence' might do [34–36]. With the term interdependence, I mean a mobile perspective in which the necessity to depend, sometimes, on others can be recognized but in which the privileged perspective is neither passivity nor autarchy.

Similar to vulnerability, interdependence is first understood as having a *universal* nature. Interdependence characterizes every aspect of human and nonhuman life (in this latter case, it can also be described as symbiosis or sympoiesis) [37]. Disability studies, as Judith Butler highlights, have remarked how every action and movement is dependent on, and is facilitated by, all sorts of human and nonhuman companions [38]. Several members of society contribute to the weaving of our lives, and "infrastructural conditions" can cause "precarity", "threat", or favour opportunities and agency [13, pp. 14–19]. Disability studies underline the leaking and unstable nature of subjects and their bodies. We, all, are captured in transversal assemblages [12, 39].

Through the lens of interdependence, it is possible to frame several aspects of disabled people's lives, for example, the need for care, support and assistance. The concept of interdependence does not ascribe passivity to disabled people and does not draw attention to their supposed incapacity to be independent. Rather, it can reveal that every subject contributes to social kinships, even when this role is not recognized. This concept not only underscores the fact that no one is capable of doing everything in autonomy but also that the areas and the degree of someone's need for support vary, and in some cases are deeper (and this can be the case with new-borns, elderly people, and people with disabilities or illnesses). Through the concept of interdependence, it is possible to understand that it is more fruitful to insist on the interrelational nature of every life, while at the same time, disabled people, like anyone else, must be guaranteed the possibility of reaching personal goals and carrying out daily activities. As claimed by the poet and activist Eli Clare, "[p]art of claiming disability is choosing this messy, imperfect work-in-progress called interdependence" [36, p. 136]. The examples of care relationships addressed in the last section of the essay are therefore inscribed within the framework of interdependence: disability *matters*, as disabled people are especially skilled in this practice and sometimes have specific needs in terms of assistance and accessibility, but at the same time, this concept transversally refers to our cohabitation as humans.

5 Third Keyword: Resourcefulness

Framing subjects as vulnerable, as I said, can produce a disempowering perspective on the subjects addressed, according to which they always need protection and assistance. This aspect, albeit sometimes true, can make us overlook their resources, capacities and competencies. Given that disabled people are often addressed as needing assistance *because* they come up as vulnerable, the space of care is rarely considered a location in which disabled people are particularly trained and skilled. This general form of disregard is blatant both in cultural texts and in fictional representations concerning disabled people situated in contexts in which infrastructures are particularly precarious, and vulnerability is therefore transversally rampant. For example, it is frequently recalled how in speculative fictional futures—especially post-Apocalyptic ones—disabled people are depicted as burdens and as almost inevitable casualties. It is imagined that they necessarily have to succumb because they are considered from a fundamentally passivizing perspective, or it is imagined that they are left behind to die [28, 40–43]. Their expertise, inventiveness and adaptability are not taken into account. As Alice Wong recalls,

> I often wonder how disabled people will survive in a postapocalyptic world […]. What I do know is that disabled people are creatures of adaptation that design and build worlds that work for them. The skills that we have reimagining/hacking/surviving hostile ableist environments would serve us well in any dystopian future. [44]

These reflections also echo current events that have resembled, in some circumstances, almost apocalyptic contours. As it has been noted by activists and scholars since the uprising of the pandemic emergency, the loss of certain lives is alarmingly considered more acceptable or simply does not cause the same dismay [45–49]. The COVID-19 pandemic has had a significant impact on disabled people, who have been disproportionately affected not only because they presented underlying health conditions that put them at higher risk but also because of ableist narratives and practices that deeply increased their vulnerability. They have faced barriers in accessing healthcare and other essential services. Additionally, measures put in place to control the spread of the virus, such as lockdowns and social distancing, have disrupted routines and support networks that many disabled people rely on for their daily lives. Last, and most importantly, disabled people have been considered, as mentioned above, *inevitable casualties* because their lives are less valued than others. There have been cases where healthcare rationing policies have prioritized younger and healthier patients over older or disabled patients or in which they have been denied life-saving treatment or access to ventilators *because* of their disability [50–55]. In this regard, the pandemic has permitted preexisting inequalities, systemic barriers, and ableist/eugenic assumptions to surface. The increase in disabled people's vulnerability should obviously be addressed, especially as a problem of social justice.

However, another point is also raised: even if attention is given to the fact that the survival of disabled people is often not guaranteed by structural injustices that include inadequate healthcare, wealth distribution, social barriers, racism, inaccessible housing, and environmental injustice, the role that disabled people themselves play in their own well-being almost always remains in the shadows. One key philosophical claim of the present contribution is that *excessive emphasis on vulnerability can disempower disabled people*, thereby also leaving their resourcefulness unacknowledged. Their vulnerability has been enhanced by the spread of the virus, by ableism and disablism, and by the suspension or fragilization of interdependence: forms of support, care, and assistance, both paid/professional and informal, have been variously disrupted due to self-protection, policies, and inadequate welfare and healthcare systems. At the same time, however, in this moment of higher vulnerability, disabled people have experimented more than ever before collective care. It is therefore important to consider the higher risks disabled people have experienced and at the same time not to overlook their adaptability and competence.

6 Introductory Notes on Care

Before examining collective care, I will briefly consider how disabled people stand in care relationships, broadly speaking. As argued above, disabled people are sometimes assumed to be more vulnerable than nondisabled people because they are particularly dependent. This aspect is especially linked to care and assistance relationships. Susan Dodds clarifies this point:

> Dependence is one form of vulnerability. Dependence is vulnerability that requires the support of a specific person (or people)–that is, care. To be dependent is to be in circumstances in which one must rely on the care of other individuals to access, provide or secure (one or more of) one's needs and promote and support the development of one's autonomy or agency. [56, p. 182]

Disability studies help to understand that the dependence inscribed within assistive relationships is not exclusively an expression of vulnerability, per se: the *quality* of these same relationships, on the other hand, can both alleviate and intensify vulnerability. Subjects can experience vulnerability because they are receiving *inadequate care*. Care relationships can also *reduce* the vulnerability experienced by a subject in a given historical, geopolitical, economic, and sociocultural context. I will focus precisely on this possibility.

Relationships between disabled individuals and those who provide care can be characterized by power imbalances and forms of (often intersecting) discrimination: sexism, fatphobia, racism, homophobia, transphobia, and even ableism. Disabled people can experience violence and harassment and endure a lack of control over their lives: care, also in its institutionalized forms, can represent "a place of oppression, loss of empowerment, physical and sexual abuse, and neglect" [28, 34, p. 3, 35, 36, p. 136]. Whereas feminist theory is principally focussed on whoever provides care work, disability scholars pay close attention to the perspective of the care receiver as well—also pinpointing how these roles are not always so clear.

My approach to care is rooted in the perspective of disabled people, who will appear both as caregivers and as care receivers, almost always simultaneously. The concept of "cure", Eli Clare argues, "requires damage": something wrong has to be found [57]. Care, on the other hand, even though possibly inscribed into asymmetrical relationships, is a more flexible category that recalls a movement, a flux, rather than cartographic attention to individual "deficits". Care is a flexible flux of emotions, actions, attitudes, practices, and narratives and can be both liberating and oppressive. Rachel Adams underlines the multiplicity inscribed within the concept:

> Care is work, an attitude toward others, and an ethical ideal. I define it as the intimate and necessary labor required to sustain those who are dependent but also the action needed to sustain the lives of vulnerable others. [58, p. 695]

I will emphasize the generative aspects and creative possibilities entailed by relationships that revolve around care but also the uncomfortable positions and the power imbalances that emerge. Disabled communities *mobilize* the nature of care, providing insights into both its necessity on a social base and the complexity that comes with it. In the last section of this essay, care crafted by disabled people *for* disabled people is framed as "a tool with which marginalized communities activate, engage in, and sustain social justice fights" [59, p. 6].

7 A Case Study: Care Collectives

I will take into consideration voluntary relationships in which the disabled participants play multifaceted roles. In collective care networks, disabled people provide physical and emotional support for one another. Disabled people "have always had to improvise when care is lacking", as Akemi Nishida underlines [59, p. 6]: discrimination and abuse that sometimes characterizes care relationships and the structural inadequacy of welfare and healthcare have forced them to experiment with "alternative support structures" compared to the paid healthcare professional or related to the family environment [28]. "Care collectives", in brief, are a way to "actively practice interdependence—a principle that disability justice activism advocates for" [28, 55, 59, p. 8, 60, 61].

Disabled people explore forms of community-based collective care for different and often intersecting reasons, already mentioned throughout the essay: because they have experienced abuse, neglect and discrimination in more "traditional" forms of care, namely, by assistance staff and biological family; because they want to establish a more profound connection with other disabled people, based on political affinities too; because they have to obviate inadequate statal support systems and low-incomes force them to do so; because they want to emancipate from family care, and so on. While similar forms of collective care have always been practised, especially within marginalized communities (queer, Black, etc.) [28, 55, 59], I am referring here to examples situated in recent years that have sometimes been accelerated by the COVID-19 pandemic and the urgency to mend the deterioration of support nets and social exchanges.

> In whatever ways they came to experience care collectives, these disabled people were eager to collectively endeavor to create different and more-just ways to meet their care needs and honor their caring capacities. [59, p. 127]

Artist and scholar Loree Erickson has been one of the first to launch—in theory and practice—a form of care network, which partly differs from the one I will present later on. In the last twenty years at least, most of her care has been provided by a rotating collective made of friends or other community members (this last case especially concerns her travel needs whenever she moves away from the usual network). This form of care is therefore based on voluntary participation, by disabled or nondisabled people, scheduled in shifts that cover Erickson's daily needs. Beyond the more settled network of people, Erickson usually posts her needs on social networks, asking to participate in her care and assistance [62, 63].

Writer and activist Leah Lakshmi Piepzna-Samarasinha grapples with collective care as well; she aims to build assistance networks that are not based on "charity" but rather on "solidarity", "mutual aid", and "respect". These concepts, in her perspective, do not "connote moral superiority of the giver over the receiver" [28, pp. 40–42]. In the "collective care" experiments recalled in her dense essay *Care work* (2018) [28] and then depicted again in *The future is disabled* (2022) [55], participants are mostly disabled, neurodivergent, sick, Mad BIPOC (Black, Indigenous and People of Color) and queer people. There are nondisabled allies helping, but these networks are actively handled by disabled people and structured upon their needs and expertise. This structure also functions as a reminder that disabled people can practice relations of care and not only passively receive them. In each network, participants set an agenda and organize diverse activities based on their needs, which may include both physical and emotional support. For example, people help each other in daily tasks (shopping, cooking, dishing, personal hygiene); share knowledge and skills on medicines and assistive technologies; promote fundraising campaigns to help a member in economic distress; elaborate reflections on Disability Justice; plan visits to members temporarily in hospital, rehabilitation, psychiatric wards; and provide emotional care, verbal or not. These experiments found diffusion in Canada and the USA, with a variety of expressions [28, 55, 59, 64, 65].

Care collectives certainly represent a survival strategy but also open up to profoundly political terrains. Whoever participates necessarily collides with disabled people's advocacy claims and checks out their needs first-hand. Therefore, there are exchanges of theoretical, political, and practical knowledge that centre on the modality of existence and the activism by the people assisted [63]. Consequently, these networks can represent a platform to inform nondisabled people about social changes advocated by disabled people. They also make it possible to craft stronger and more conscious connections *within* the disability community that originate from everyone's needs and perspectives. The aim is not only to support each other but also to guarantee transversal accessibility. Care can actually represent a space where it is possible to realize "community building" and—unlike the most common narratives of it—can also be "a site of pleasure" and joy [28, p. 41, 56, 136–137, 55, 58]. Furthermore, practising and sharing these experiences contribute to challenging assumptions of disabled people's passivity and incompetence: they can in fact organize complex and stratified care networks, contributing in heterogeneous ways [65]. They are a form of resistance to narratives about ableism—a system deeply entrenched with capitalism—which rewards self-sufficient and productive bodyminds.

Distribution of care sometimes emerges as critical, mirroring society at large: in care networks, most labour is often "assigned" to specific sections of the population: women and BIPOC. It is therefore necessary to open uncomfortable conversations about care load [28, 55, 59, 64]. Members must learn to set boundaries and respect their mental and physical energy. Within this framework, everyone shall contribute starting from their possibilities, bodies, and relational inclinations. That does not mean, however, that reciprocity within these care webs must necessarily actualize in an even contribution by all members—also because some of them can be difficult to weigh. Is it possible to ponder the support in the writing of a business e-mail, giving information on disability politics, and transferring someone from the bed to the bathroom? This kind of mutuality, therefore, always comes into being starting from a plurality of support activities. These relationships entail exchanges on several levels—for example, affective ones. In Erickson's case, "logistics" that concerns her necessity is interlaced by her giving relationship advice, small talks with assistants, and general conversations about personal and popular matters [36, p. 136].

When the COVID-19 pandemic hit, disabled people found themselves in a terrifying place not only because of the virus itself but also because of forced isolation and the collapse of many webs of support. Since collective care was already an ongoing experiment—it is never a settled track—they already had some tools to organize and make survival possible *without leaving anyone behind* [28, 55, 64, 66].

> All the wisdom of disability and intersecting marginalized communities who are always and already surviving crisis after crisis without support from the government, public, and private entities but with each other have been illuminated. [59, p. 182]

Already running interpersonal connections, social media platforms, virtual communities, and local mutual-aid groups, continued to share information about the pandemic, and to support each other on different levels, often check-in in case companions and friends were isolated and anxious for their lives, in pain, sick, needing help [55, 59]. Not only did *they keep each other alive* [55], but they worked together to decrease the vulnerability collectively experienced, mostly by disabled and other marginalized people. They found themselves navigating in a frightening space—but not so unfamiliar. "Crip wisdom" and disabled skills became of primary importance, also beyond collective care—the "disabled people who were supposed to be the first to die, who absolutely refused to be sidelined or forgotten about". All the care work went in parallel, however, with "grief and terror and uncertainty and loss", which must be acknowledged as well as an outcome of the pandemic [55, p. 36]. The patterned, complex and coparticipated nature of the moment is described as such by Clare:

> We who check in
>
>   every day over text, phone, Zoom, Skype,

Facebook, FaceTime: *how are your lungs,*
can you make rent this month, did you lose your job today,
are you hungry right now, do you have enough
insulin, estrogen, Prozac, Klonopin, blood pressure meds?
We who drive across town to deliver saltines,
fresh kale, chicken soup, half bottles
of Tylenol, the last box of face masks
to ex-lovers and best friends. We who have always
shared everything we had. We who keep
each other alive. We who will be turned away
from emergency rooms and denied
ventilators. We who will never
go to the hospital. We who will die
and we who will live. [67]

As Piepzna-Samarasinha argues, growing mutual aid discourses, from 2020 onwards, seemed not to acknowledge that disabled people were *already* doing this same work for years: since the "pandemic is a disabled event", it should be clear that the disabled community may have methods, practices and insights to share [55, p. 63].

8 Conclusion

After having addressed the stratified connections between disability, vulnerability, and (inter)dependence, I have proposed a case study, care collectives, in which disabled people have actively engaged with the contextual vulnerability that can emerge in "traditional" care relationships, or within welfare and healthcare framework which reveals adequate—as it has especially occurred through the COVID-19 pandemic. In these cases, disabled people's lives are not simply rendered harder: their very survival is at risk. As emerged, challenges posed by vulnerable presents and futures may be addressed thanks to the contribution of diverse subjects, as anyone might be a bearer of valuable contributions—and not *despite* disabilities. Relational and survival practices require the expertise of people with heterogeneous cognitive, bodily, and sensory traits. Even in broken worlds, crises, and emergencies, the lessons they can share should be cornerstones in community organizing: disabled people are experts crafting assemblages centred on care and support. Furthermore, these assemblages directly intervene in social and cultural narratives of disability—for example, disabled people are assumed to be merely passive.

Disabled people highlight the interdependence of our lives, with all the opportunities and the limits entailed: since they are frequently more *exposed* and more *extroflected* beyond their individual boundaries, they are also productively engaged with others. I have highlighted the generative potential of this shared dimension:

overcoming disability as an individual experience can benefit disabled people. In the mentioned cases, experiences of illness and disability are not isolated and detached from the relational context but are shared, both emotionally and practically. Disability can become a place of emotional, affective, and care-sharing and can also represent an important piece in building a disabled culture, with knowledge and practices mutualised. The situational vulnerability they may experience can also become, in Butler's words, "a potentially effective mobilizing force" [13, p. 14]. In this sense, vulnerability on the one hand and resistance and action on the other hand do not necessarily diverge but coconstruct each other. Connections and relationships involved in one's experience of disability do not exclusively produce a deepening of dependence (and therefore of possible vulnerability) but can instead function as "supports that shield against or minimize exposure to risks" [18, p. 214]. "Misfitting", as Garland-Thomson argues, can also enhance resourcefulness: an inadequate relationship between the subject and the context (material, social, and so on) can push her/him to imagine a new reality and to negotiate more just and fertile "juxtapositions" [33]. It is important to underline that I do not aim to romanticize vulnerable ecologies but acknowledge that subjects (forced to be) at the margins produce valuable knowledge and must not be cast aside in crises either.

As COVID-19 has been a "mass disabling event", the knowledge of disabled people should be recalled as especially valuable to manage the individual and collective consequences and changes. As Piepza-Samarasinha recalls, "the pandemic has meant that there's a huge wellspring of disabled culture, collective care, communities, love, grief work, joy": this expertise should be valorised on a social level, as "a cripping of the world, more than ever before", is happening [55, p. 325]. It is urgent to incorporate this training within communities: the role of collective care networks might increase in the future because we will probably live longer, state resources may decrease, climate change and pandemics will continue to have disabling effects, and not all disabilities are recognized to receive welfare aid anyway. Collective care has mitigated the exposure to vulnerability in recent years and during the pandemic as well but can also decrease our experience of it as an ontological human condition, as we are all more likely to thrive in interdependence rather than in isolation.

However, it is important to underline once again how these examples also mirror the inadequacy and scarcity of resources that characterized our society, especially when it confronts disabled people: the forms of "resistance" and organizing presented often work as an adjustment of structural lacks, violence, and discrimination. In addition, the emphasis on disabled people's contributions in this sense cannot become a form of social exoneration: it should not turn into a comforting narrative that implies the conservation of the status quo. Disabled people must not be ceaselessly fragilized by the economic, medical, environmental and social context. Therefore, the acknowledgement of their practices must also indicate what is missing on a community level. Last, these skills and knowledge should not be *extracted* from the disabled community and appropriated: the source must always be clear. In the diverse community they imagine and practice, disabled people are not at the margins: they not only actively participate in transformative social change but also guide it.

In conclusion, throughout the essay, I have attempted to show a set of diverse possibilities of what vulnerable positionings may entail. To gain deep insights concerning disability, vulnerability, and care, it is essential to include a *crip* perspective that originates from disabled people's expertise and knowledge and is also willing to destabilize rigid assumptions about all three concepts. As these concepts are often grappled with each other, it seems important to experiment with diversified approaches whenever we address them. Keeping in mind all the cautions and the ambivalences presented in the essay, I hope a multifaceted analysis of their intersections can operate as a starting point to spark several other conversations that place disabled people's experiences at the centre.

— In vulnerable ecologies, disabled people not only are not disposable but also represent a collective resource.

Core Messages

- Vulnerability and disability are not inherently coextensive, but an intersecting analysis can be productive to uncover the relational nature of both concepts.
- Assuming that every human being is vulnerable for her/his perishable, embodied and embedded nature, it is however risky to frame as vulnerable specific subjects. It appears more fruitful considering vulnerable ecologies, social positionings and processual experiences instead.
- The inherent interdependence of our lives can both be a source of empowerment and deepen the experience of vulnerability of the subjects involved in every relational network.
- Disabled people are not exclusively care receivers, but are experts in organizing and managing care, as it emerges for example in care collectives.
- Care relationships, and the space of assistance and (inter)dependence, can be inhabited with joy, political tensions and love for the disabled community.
- Since care is a fraught and ambivalent space, and we will probably need care more than ever (because of longer lives, climate change, pandemics, etc.), it is urgent to valorise disabled people's skills and experiences.

References

1. Butler J (2004) Precarious life: the powers of mourning and violence. Verso, London
2. Butler J (2009) Frames of war: when is life grievable? Verso, London
3. Held V (1987) Non-contractual society. In: Hanen M, Nielsen K (eds) Science, morality and feminist theory. University of Calgary Press, Calgary, pp 111–137
4. Feder-Kittay E (1999) Love's labor: essays on women, equality, and dependency. Routledge, New York
5. Mackenzie C et al (eds) (2014) Vulnerability: new essays in ethics and feminist philosophy. Oxford University Press, Oxford

6. Koivunen A et al (2018) Vulnerability as a political language. In: Koivunen A, Kyrölä K, Ryberg I (eds) The power of vulnerability. Manchester University Press, Manchester, pp 1–26
7. Butler J et al (2016) Introduction. In: Butler J, Gambetti Z, Sabsay L (eds) Vulnerability in resistance. Duke University Press, Durham, pp 1–11
8. Shildrick M (2002) Embodying the monster. Encounters with the vulnerable self. SAGE, London
9. Shakespeare T (2014) Disability rights and wrongs. Routledge, New York
10. Rohrer J (2011) Toward a full-inclusion feminism: a feminist deployment of disability analysis. Fem Stud 31(1):34–63
11. Moser I (2020) Against normalisation: subverting norms of ability and disability. Sci Cult 9(2):201–240
12. Shildrick M (2015) "Why should our bodies end at skin?", Embodiment, boundaries, and somatechnics. Hypatia 30(1):13–29
13. Butler J (2016) Rethinking vulnerability and resistance. In: Butler J, Gambetti Z, Sabsay L (eds) Vulnerability in resistance. Duke University Press, Durham, pp 12–27
14. Casalini B (2018) Le teorie femministe contemporanee, dal paradigma della sovranità al paradigma della vulnerabilità. Bernardini et al, a cura di, Vulnerabilità: etica, politica, diritto. IF Press, Roma, pp 27–48
15. Ahlvik-Harju C (2016) Disturbing bodies: reimagining comforting narratives of embodiment through feminist disability studies. Scand J Disabil Res 18(3):222–233
16. Cavarero A (2009) Inclinations: a critique of rectitude. Stanford University Press, Redwood City
17. Wendell S (1997) The rejected body. Feminist philosophical reflections on disability. Routledge, New York and London
18. Scully JL (2014) Disability and vulnerability: on bodies, dependence, and power. In: Mackenzie C et al (eds) Vulnerability: new essays in ethics and feminist philosophy. Oxford University Press, Oxford, pp 204–221
19. Shildrick M (2009) Dangerous discourses of disability, subjectivity and sexuality. Palgrave Macmillan, New York
20. Butler J et al (eds) (2016) Vulnerability in resistance. Duke University Press, Durham
21. Wong A (2017) My medicaid, my life. https://www.nytimes.com/2017/05/03/opinion/my-medicaid-my-life.html
22. Bernardini MG (2018) Disabilità, vulnerabilità e diritti (umani). Tra conflitti e (nuove) opportunità. Ead et al, a cura di, Vulnerabilità: etica, politica, diritto. IF Press, Roma, pp 211–236
23. Kafer A (2013) Feminist, queer, crip. Indiana University Press, Bloomington
24. Goodley D (2011) Disability studies. An interdisciplinary introduction. SAGE, London
25. Oliver M (1990) The politics of disablement. Palgrave Macmillan, New York
26. Barnes E (2016) The minority body: a theory of disability. Oxford University Press, Oxford
27. Chapman R (2020) Neurodiversity, disability, wellbeing. In: Bertilsdotter-Rosqvist H et al (eds) Neurodiversity studies. Routledge, London and New York, pp 57–72
28. Piepzna-Samarasinha LL (2018) Care work. Dreaming disability justice. Arsenal Pulp, Vancouver
29. Kleinman A, Kleinman J (1991) Suffering and its professional transformation. Cult Med Psychiatry 15(3):275–301
30. Clare E (2015) Exile and pride: disability, queerness, and liberation. South End, Boston
31. Montalti C (2021) Vulnerabilità 'tecnogena' e condivisione tecnologica nell'esperienza della disabilità. Shift. Int J Philos Stud 8(2):159–171
32. Montalti C (forthcoming) Neurodivergence through Octavia E. Butler's speculative futures: community, vulnerability, and social change. In: Creechan L, Bergenmar J, Stenning A (eds) (Neuro)divergent textualities. Representations, readings, responses. Bloomsbury, London
33. Garland-Thomson R (2011) Misfits: a feminist materialist disability concept. Hypatia 26(3):591–609

34. Fritsch K (2010) Intimate assemblages: disability, intercorporeality, and the labour of attendant care. Crit Disabil Discourses 2. https://cdd.journals.yorku.ca/index.php/cdd/article/view/23854
35. Baggs M (2012–2013) https://ballastexistenz.wordpress.com/tag/caregivers/
36. Clare E (2017) Brilliant imperfection. Grappling with cure. Duke University Press, Durham and London
37. Haraway DJ (2016) Staying with the trouble. Making kin in the Chthulucene. Duke University Press, Durham and London
38. Butler J (2017) Notes toward a performative theory of assembly. Harvard University Press, Cambridge and London
39. Shildrick M (2010) Some reflections on the socio-cultural and bioscientific limits of bodily integrity. Body Soc 16(3):11–22
40. Allan K, al-Ayad D (eds) (2015) Accessing the future. Futurefire.net
41. Durrow R, Berne P (2020) We love like barnacles. Into the crip universe: cripping the anthropocene (podcast)
42. Duyvis C (2016) Fiction, like life, tells disabled people their lives have less value. The Guardian. https://www.theguardian.com/childrens-books-site/2016/mar/14/disability-autism-sci-fi-corinne-duyvis
43. Weise J et al (2019) Cyborgs. Disability visibility project
44. Wong A (2019) The rise and fall of the plastic straw. Catal Feminism Theory Technoscience 5(1):2
45. Taylor S, Orning SES (2020) Being human, being animal: species membership in extraordinary times. New Literary History 51(4):663–685
46. Kukla E (2020) My life is more 'disposable' during this pandemic. The New York Times. https://www.nytimes.com/2020/03/19/opinion/coronavirus-disabled-health-care.html
47. Ware SM, Clare E (2020) Disabled: not a burden, not disposable. Transgender Law Center. https://www.youtube.com/watch?v=jA8PkucFrCk&ab_channel=TransgenderLawCenter
48. Mingus M (2022) You are not entitled to our deaths: COVID, abled supremacy & interdependence. Leaving Evidence. https://leavingevidence.wordpress.com/2022/01/16/you-are-not-entitled-to-our-deaths-covid-abled-supremacy-interdependence/
49. Tremain S (ed) (2021) Philosophies of disability and the global pandemic [special issue]. Int J Crit Divers Stud 4(2)
50. Human Rights Watch (2021) Disability rights during the pandemic. https://www.hrw.org/report/2021/03/24/disability-rights-during-pandemic/inequality-dignity-and-access-life-saving
51. United Nations Human Rights Office of the High Commissioner (2020) Covid-19 and the rights of persons with disabilities: guidance. https://www.ohchr.org/Documents/Issues/Dis-ability/COVID-19_and_The_Rights_of_Persons_with_Disabilities.pdf
52. Alfonseca K (2022) Disabled people being left behind in vaccine rollout one year later. ABC News. https://abcnews.go.com/Health/disabled-community-left-vaccine-rollout-year/story?id=81974345
53. Purlang A (2022) 6 ways responses to covid-19 have been ableist, and why it matters. Forbes. https://www.forbes.com/sites/andrewpulrang/2022/01/31/6-ways-responses-to-covid-19-have-been-ableist-and-why-it-matters/?sh=6203060a6129
54. Shapiro J (2020) One man's COVID-19 death raises the worst fears of many people with disabilities. NPR. https://www.npr.org/2020/07/31/896882268/one-mans-covid-19-death-raises-the-worst-fears-of-many-people-with-disabilities
55. Piepzna-Samarasinha LL (2022) The future is disabled. Arsenal Pulp, Vancouver
56. Dodds S (2014) Dependence, care, and vulnerability. In: Mackenzie C et al (eds) Vulnerability: new essays in ethics and feminist philosophy. Oxford University Press, Oxford, pp 181–203
57. Clare E (2016) Disability, queerness, and natural worlds. https://www.swarthmore.edu/news-events/listen-author-eli-clare-cure-disability-queerness-and-natural-worlds
58. Adams R (2020) The art of interspecies care. New Literary History 51(4):695–716
59. Nishida A (2022) Just care. Messy entanglements of disability, dependency, and desire. Temple University Press, Philadelphia

60. Kafai S (2022) Crip kinship: the disability justice & art activism of sins invalid. Arsenal Pulp, Vancouver
61. Mingus M (2017) Access intimacy, interdependence and disability justice. Leaving Evidence. https://leavingevidence.wordpress.com/2017/04/12/access-intimacy-interdependence-and-disability-justice/
62. Erickson L (2018) Welcome to the new ELA post-doctoral fellow. Vision Passion Action. https://radssite.wordpress.com/2018/09/24/welcome-to-the-new-ela-post-doctoral-fellow-dr-loree-erickson/
63. Hande MJ, Mire M (2013) "The pace we need to go": creating care culture. Action Speaks Louder, 10–11
64. https://creatingcollectiveaccess.wordpress.com/
65. Brown LXZ (2017) Ableist shame and disruptive bodies: survivorship at the intersection of queer, trans, and disabled existence. In: Johnson AJ et al (eds) Religion, disability, and interpersonal violence. Springer, Berlin, pp 163–178
66. Sins Invalid (2015) 10 principles of disability justice. https://www.sinsinvalid.org/blog/10-principles-of-disability-justice
67. Clare E (2021) May Day, 2020. The South Atlantic Quarterly 120(2):255–256

Chiara Montalti (1993) is a Postdoctoral researcher in Moral Philosophy at the University of Salento. She obtained her Ph.D. in Philosophy at the Joint Florence-Pisa Universities Doctoral School in 2022, with a thesis in Disability Studies and Feminist Posthumanism and Technoscience. Her main research interests are technological assemblages, care relationships, disability representations in fictional worlds, and political activism. In 2021 she has conducted a thesis-related research period in collaboration with the University of Oslo. In 2021 and 2022 she collaborated, in a multimedia philosophical project and in communication, respectively, with the Istituto Italiano per gli Studi Filosofici.

Chapter 11
"Total Institutions" as Litmus Test of Civilisation

Chantal Marazia, Nils Löffelbein, and Heiner Fangerau

Abolish prisons!
Thomas Galli, Die Zeit, 14. May 2020

Abstract The COVID-19 pandemic has exacerbated and magnified the vulnerability of convicts, psychiatric patients, elderly and disabled people in need of care, and children in institutional care or custody. Taking the move from the media coverage of "Total Institutions" during the pandemic, this paper explores the ways in which Goffman's concept was mobilised in the public discourse to raise awareness of the extreme predicament of these vulnerable groups, as well as of the whole locked-down society. Whereas the metaphoric expansion of this *terminus technicus* to cover the whole community under lockdown threatens to dilute its specific meaning and overshadow the special vulnerability of inmates, we argue that the feeling of shared experience of "totalisation" bears a transformative potential. The possibility of taking stock of the multilayeredness (cognitive, rational, emotional) of the shared experience may enable collective cognizance of the structural roots of special vulnerability, thriving in the extant total institutions.

Keywords "Total institutions" · Confinement · Goffman · COVID-19 · Vulnerability · Media coverage

1 Introduction

Long before the outbreak of the COVID-19 pandemic, former prison director Thomas Galli repeatedly called for a fundamental reform of the (German) penal system, the core of which included the abolition of prisons [1, 2]. His militancy, however, had been confined to scholarly, professional and militant circles, without a chance to

C. Marazia (✉) · N. Löffelbein · H. Fangerau
Department of the History, Philosophy and Ethics of Medicine, Medical Faculty, Heinrich-Heine-Universität Düsseldorf, Düsseldorf, Germany
e-mail: marazia@hhu.de

S. Achella and C. Marazia (eds.), *Vulnerabilities*, Integrated Science 18,
https://doi.org/10.1007/978-3-031-39378-5_11

reach the wider public. In May 2020, his voice was eventually given a prestigious tribune, in the popular German weekly *Die Zeit* [3]. In the following months, the evolution of the pandemic scenario provided a specific framework of reference to his otherwise more general argument [4, 5].

The fate of Galli's plea is not unique: The combination of a viral threat and increasing restrictions within prisons has caused anti-carceral advocates and civil society representatives to challenge correctional systems across the globe [6, 7]. Only four days after the proclamation of the pandemic, the World Health Organization warned that "[p]eople deprived of their liberty [...] are likely to be more vulnerable to the coronavirus disease (COVID-19) outbreak than the general population because of the confined conditions in which they live together for prolonged periods of time. Moreover, experience shows that prisons, jails and similar settings where people are gathered in close proximity [such as detention centres and refugee camps] may act as a source of infection, amplification and spread of infectious diseases within and beyond prisons" [8]. According to a survey published by the organisation Human Rights Watch, within a couple of weeks, almost every second country announced 'back end' measures to reduce the spread of COVID-19 behind bars: At least 80 governments authorised the release of more than 580,000 detainees [9]. However, most announcements remained dead letter. In Europe, for instance, only 16 countries adopted this policy [10]. The majority of correctional facilities introduced 'front end' and 'in prison' mitigation strategies: suspension of social and family contact, as well as of temporary leave, and reduction of daily movement within the institutes [11, 12].

In the very first months of the COVID-19 pandemic, prisons never left the limelight—from the deadly riots in Italy and Brazil [13] to the hunger strikes in the USA. The *New York Times* even opened a forum for the narratives of incarcerated people [14]. However, prisoners and people in refugee camps are not the only groups clustered together in (unhealthy and) disease-spreading environments, whose vulnerability has been both exacerbated and magnified by the latest pandemic [15].[1] Psychiatric patients, elderly and disabled people in need of care, and children in institutional care or custody are also considered particularly vulnerable. Many of them are consigned to what Erving Goffman famously called "total institutions" in his 1961 collection of essays *Asylums* [16].[2] Although Goffman never used the term "vulnerable", the majority of these institutions are almost by definition receptacles of

[1] In their analysis of press across the United States, Schneeweis and Foss found ample media coverage in 2020, but a "striking absence and ignorance of key developments later into 2021". During this timespan, they identified six different yet interconnected discourses: "journalistic objectivity, blaming and abandonment, vulnerability, compassion, vilification, and absence" [15, p. 187].16

[2] Goffman censed a vast number of "total institutions", classed into five types according to function: institutions conceived in order to care for those who cannot care for themselves (e.g. orphanages, homes for the elderly or the blind); institutions designed as a sanctuary for those who voluntarily retreat from the world (e.g. convents and monasteries); institutions conceived for the care for those who can not care for themselves and may constitute an unintentional threat to the community (such as psychiatric patients or those with an infectious disease); institutions conceived in order to protect the community against what are perceived as intentional dangers (e.g. prisons, jails, prisoner of war and concentration camps); institutions established in order to achieve an educational or work task (e.g. army barracks, boarding schools, work camps or ships) [see].

marginalised and already-vulnerable individuals. Many of these facilities turned out to be "lethal weapons" [17], "the ground zero of the COVID-19 pandemic" [18], and their inmates gained dramatic media visibility at the outbreak of the sanitary emergency. Through their tragic bulletins of casualties, "ugly stepchild[ren] of health care" [19, p. e214] such as nursing homes and long-term facilities became "a leading story" almost overnight [19, p. e214].

At times of global crisis, such as pandemics, news reporting captures information reflecting and informing citizens' experiences. It also ignites debates on the effects of public health policies, social responsibility and ethics and is therefore crucial to understanding the choices of policy makers, the role of experts and the concerns of the public. As historians with an interest in "total institutions", we were struck by this media coverage, the more so, as it stood in sharp contrast with the first spur-of-the-moment attempts at diagnosing the state and possible futures of our society through the lens of the emergency and the measures it called for—both perceived as "unprecedented". The many learned ponderations over this nexus exclusively focused on the disquietingly sudden and smooth reconfiguration of the daily life of ordinary people into a biopolitical experiment [20]. Renowned intellectuals did not spare historical analogies, but removed them quite ahistorically from the pandemic context. The theorist of the "state of exception", Giorgio Agamben, directly compared the state's measures to the Nazis' repression of Jews; the philosopher pictured Italy as being on a direct path to dictatorship [21]. With similar references to twentieth century totalitarianism, German philosopher Peter Sloterdijk likened state restrictions to the 1933 "Machtergreifung" (seizure of power) in an interview with the French newspaper *Le Point* [22].

The impression one gathered from the first debates was that only the extraordinary measures affecting free and autonomous individuals are truly symptomatic of our degree of civilisation—a diagnosis paradoxically oblivious of the classic spaces of biopower, such as prisons, psychiatric hospitals, and children's homes. The debate revolved around questions such as "is such a heavy administrative intervention really justified?", or "what is the limit beyond which we are not prepared to renounce the political and ethical principles shaping our existence?" The focus here was on an (alleged) loss of well-established values and inalienable principles.

In a contribution to this very debate, however, Daniele Lorenzini has dutifully reminded that "biopolitics is always a politics of *differential vulnerability*. Far from being a politics that erases social and racial inequalities by reminding us of our common belonging to the same biological species, it is a politics that structurally relies on the establishment of hierarchies in the value of lives, producing and multiplying vulnerability as a means of governing people" [23]. If, moreover, one agrees that "total institutions" and their political and societal management are the litmus test of a civilisation, their conspicuousness can be an appropriate point of departure for a diagnosis of our society as well. Starting from places that are by definition marginal and closed, in this chapter we broaden the perspective, in search of principles on which our society may rely in the future. Our first questions, then, are as follows: How come such places of marginality par excellence gained such a mediatic centrality in this conjuncture? Can it be considered a feature common to major sanitary emergencies, or was this specific to this contingency?

2 "Total Institutions" During Pandemics: A Historiographical Desideratum

The most adequate way of addressing these questions would be a comparative historical analysis of the impact of major epidemics and pandemics on the contingent management and media coverage of "total institutions" and their inmates. More than half a century before Goffman's *Aslyums*, German historian and pacifist Ludwig Quidde published a series of articles in the *Münchener Freie Presse* about "Poor people in the hospital" [*Arme Leute in Krankenhäusern*], denouncing loss of autonomy and abuse of the poor in institutional, clinical settings. His focus, however, was not on internment itself, but on the exploitation of patients in research without their consent [24]. In general, a loss of individual autonomy in medical settings often results from facts defined by biologically grounded criteria. Public health actions are de facto based on biological arguments. These biologically interpreted bases for action are again shaped according to the prevailing interpretations of health and disease in a particular civilisation. The interpretation of what is recognised as a disorder, the techniques adopted for dealing with it, the individual perception of suffering and its public perception are historically variable and culturally specific [25, p. 122].

Several studies have already considered single facilities during epidemic events, how they were impacted and how they reacted as self-contained systems, showing that restrictive measures such as those adopted during the COVID pandemic are not new. For instance, it is a known fact that the two (now almost forgotten) great waves of influenza of 1957/58 and 1968/69, which killed an estimated 70,000 people in West Germany alone, disproportionately hit prisons, nursing homes and psychiatric hospitals throughout Europe. Due to the rapidly increasing number of infections among patients and staff, in December 1968, the Ludwigsburg District Hospital immediately banned visitors. During the polio epidemics of 1953/54, which caused 10,000 deaths in the Federal German Republic only, children's homes and similar institutions were closed down, as the disease mostly affected children and the young [26]. However, remarkably little attention has been devoted to structural issues, such as the recognition and definition of the (exceptional) vulnerability of prisoners and patients during sanitary crises in the public sphere.

In her history of epidemics, spanning from the 1721 smallpox outbreak in Boston to the 1952 poliomyelitis in the United States, through the 1918–1919 "Great Influenza", Katherine Foss has shown that inmates have historically received the least public attention in times of sanitary emergencies. At different times and places, media outlets appear to show a consistent tendency to focus on threats to privileged classes, ignoring precisely those who are most vulnerable to disease [27]. On the basis of the available empirical evidence, we assume that the currency of "total institutions" in the media during the coronavirus-pandemic is indeed an unprecedented phenomenon and that something new has occurred in the moral (and political) imagination of Western societies.

3 The Academic Renaissance of "Total Institutions"

Alongside this allegedly unmatched prominence of "total institutions" as such, what we found remarkable was the renewed fortune of the Goffmannian expression, both in the scholarly literature and in essayistic columns in the press dealing with the impact of COVID-19 on Western societies. Although it rarely made it to the title [see 28, 29], the term prominently featured as a key concept in a plethora of academic studies.[3]

Looking at numbers, what catches the eye is not only the renewed fortune of the concept of total institution during the coronavirus pandemic, but also its semantic broadening. In academic writings, the expression was not limited to the original Goffmanian catalogue (mostly to prisons, nursing homes, psychiatric hospitals, or ships), but extended to facilities hardly associated with it before, such as student dorms [see 30] or zoos [see 31]. This expansion is arguably due to the fact, that some facilities (think of quarantine hotels) actually gained both total and institutional traits only during the pandemic.

Goffman's original formulation of the concept reads as follows:

A basic social arrangement in modern society is that the individual tends to sleep, play, and work in different places, with different co-participants, under different authorities, and without an overall rational plan. The central feature of total institutions can be described as a breakdown of the barriers ordinarily separating these three spheres of life. First, all aspects of life are conducted in the same place and under the same single authority. Second, each phase of the member's daily activity is carried on in the immediate company of a large batch of others, all of whom are treated alike and required to do the same thing together. Third, all phases of the day's activities are tightly scheduled, with one activity leading at a prearranged time into the next, the whole sequence of activities being imposed from above by a system of explicit formal rulings and a body of officials. Finally, the various enforced activities are brought together into a single rational plan purportedly designed to fulfil the official aim of the institution [16, pp. 5–6].

A quantitative analysis of Goffman's academic revitalisation is not the scope of this contribution. What we want to focus on here, is the use of Goffman's expression in the public discourse. In what follows, we will rely on the German case, on the assumption that it shares fundamental traits with other national realities. At the

[3] In March 2023 on Google scholar the combination of "total institutions" and "COVID" produced 853 results, 888 for "total institution" and "COVID".

same time we are aware of possible national idiosyncrasies—which may reflect the different reform paths the single countries took in the last sixty years [see].

For decades, historians of punishment and confinement have used Goffman's theory relatively unreflectively, as an explanatory foil for the specific modes of functioning and courses of action in psychiatric and correctional institutions. The underlying reasons for the pronounced structural and personal relationships of violence in inpatient facilities or penitentiaries were mostly read through this lens [35]. However, the sociological model of the "total institution" has come under criticism in recent years: researchers are now increasingly asking whether Goffman's theory should not be considered outdated by social change, calling the totality of contemporary institutions into question [36–38]. Above all, the concept would not cover the transformation of the institutional landscape reflected by the progressive deinstitutionalisation of psychiatric care [39], or by the increased "porousness" of prisons in the Western world since the 1970s [40]. Outpatient "home treatment" of Swiss psychiatry in 2022 has indeed little in common with the prison-like institutions of the 1950s and 1960s [41]. Besides, historians have disclosed that Goffman's rigid conception of roles and the strict separation between staff and "inmates" does not take into account the complex social relations structuring closed institutions far before the psychiatric reform [see 42].

4 The Total Institution in Public Perception

The social need for orientation and interpretation of far reaching changes in everyday life has grown rapidly since the outbreak of the coronavirus pandemic in spring 2020. Physical distancing, contact bans [see on this 43] and strict rules of conduct restricted individual sovereignty and fundamental rights, such as the freedom of movement and assembly of citizens, to an extent hardly considered possible before. As the newspaper *Frankfurter Allgemeine Zeitung* accurately acknowledged in an article at the end of 2020, the (first) COVID year had "inspired social science to make far-reaching diagnoses" [44]. Against this background, Goffman's model was eagerly taken up by numerous observers to explain the exercise of institutional power and, significantly, its repressive facet was given special prominence.

Above all, the introduction of strict quarantine measures in inpatient facilities such as prisons, hospitals and homes, where residents were isolated and kept apart from each other, triggered a public debate on the more or less subtle use of institutional coercion [45]. People were used to open doors in institutions such as nursing homes or hospitals. During the pandemic, the gateway, the connection between the inner and the outer world, regained its traditional function of space-related tool of inclusion and exclusion [46, pp. 116–121]. In November 2020, the educationalist Peter Rödler diagnosed a "renaissance" of closed accommodation in the wake of the COVID-19 pandemic, which clearly exhibited features of the "asylumisation" of vulnerable groups described by Goffman [47]. Swantje Köbsell, Professor of Disability Studies at the Alice Solomon University of Applied Sciences in Berlin, also argued, in an interview with the weekly newspaper *Der Freitag*, that, during the pandemic, the "institutions became truly total again" [48]. In this context, legal sociologist Johannes Feest pointed to the seamless transfer of terms, such as "contact ban" and "distance requirement", from the penal system to the state's coronavirus measures policy [49]. The social pedagogue and association functionary Klaus Heck even described nursing home residents as "passive object[s]" of the state's policy of measures and as "generally incapacitated" [50].

Frequently, the reference to Goffman's organisational model was accompanied by a general critique of society, suggesting that the weakest and most vulnerable members of society had been pushed into closed (and total) institutions by the ordinance policy, or had been deliberately "asylised". The argument of "protecting at-risk groups" was seen as no more than a pretext for a policy of segregation, a sealing off "out of convenience", as the journalist Frédéric Valin wrote in January 2021 [51, p. 6]. Precisely the highly vulnerable residents of homes, hospitals, psychiatric wards and prisons would have been "sacrificed" by society and politics. Cultural sociologist Monika Wohlrab-Sahr described as one of the "paradoxes of sociality" during the COVID-19 crisis the fact that the supposed protection of vulnerable groups in old people's and nursing homes actually served to protect the majority of society [52].

While some authors considered the hardships of the lockdowns as proximate cause of the emergence of totalitarian tendencies in institutions sealed off from the outside world, others were convinced that the pandemics had only brought to light the already existing grievances in homes, psychiatric institutions and similar types of organisations [53]. Paradoxically, the total closure of the long term facilities and nursing homes of the inpatient institutions would not have protected the people accommodated but on the contrary would have turned the institutions into real "death traps" [51, p. 6].

The exorbitant number of victims among nursing home residents throughout Europe has thus often been identified as a direct consequence of the structures, working conditions and living situations in "total institutions". However, many commentators agreed that the coronavirus-induced "state of emergency" decisively aggravated the accommodation conditions. In April 2021, the managing director of the German association "Interessensvertretung Selbstbestimmt Leben in Deutschland" (an organization representing the interests of disabled people), Rebecca Maskos, lamented the acute increase in structural violence in institutions for persons with disabilities in the wake of the first coronavirus lockdown: not only were the psychological consequences of social isolation in homes frightening; in some cases, quarantine measures were enforced with physical coercion, and residents were sedated with medication and locked away in their rooms for long periods of time [54].

The extent to which coercive measures in inpatient placement are ethically and legally justified in the epidemiological state of emergency, has also been matter of wide scholarly debate. Here, too, the model of the "total institution" was used as a reference in numerous works [55]. As early as in March 2020, the European Committee for the Prevention of Torture and Inhuman or Degrading Treatment or Punishment issued a policy statement on the treatment of persons deprived of their liberty, i.e. persons in police detention facilities, penitentiary institutions, immigration detention centres, psychiatric hospitals and social care homes, as well as in newly established quarantine spaces. While pointing out the clear imperative to take firm measures to combat the virus, the Committee reminded "all actors of the absolute nature of the prohibition of torture and inhuman or degrading treatment. Protective measures must never result in inhuman or degrading treatment of persons deprived of their liberty" [56].

5 Political Agenda Setting

Often, the evocation of "total" conditions in closed institutions was also functional in advancing policy-related concerns. In February 2021, the Federal Association for the Assistance of Offenders (*Bundesarbeitsgemeinschaft für Straffälligenhilfe e.V.*), while praising "the start of the vaccination campaign in old people's homes and nursing homes", called for "appropriate priority to be given to prisoners and staff in total institutions, i.e., prisons and prison-like institutions" [57].

Many actors subsumed a whole series of criticisms under Goffman's catchy term, most of which had already been formulated by representatives of the deinstitutionalisation movement long before the outbreak of the pandemic. In this respect, the "total institution" also functioned as a cipher to address traditional structural problems of the nursing, hospital and justice sectors in a particularly pointed and effective way. In the early 2000s, the interdisciplinary research working group "Menschen in Heimen" (People in Homes/Asylums) at the University of Bielefeld pleaded for the establishment of an "Enquete der Heime" (Enquete of Homes/Asylums), modelled on the 1975 "Psychiatrie Enquete", to vouch for the deinstitutionalisation of people with disabilities [58]. Such evocative expression was in itself a political statement, meant to underscore the structural nature of the coercion, violence and neglect experienced by the inmates [59]. The pandemic scenario, with its exacerbated contention measures and the wide limitation of individual liberties, revived these already established radical stances, giving them a new centrality and urgency, as well as an audience they could not possibly achieve in "normal" times. Comprehensive "protection of inmates of such total institutions against totalitarian attacks" could only be ensured by abolition of "total institutions" themselves, argued social education worker Klaus Heck in an article at the beginning of the pandemic, as "is currently becoming unmistakably clear how little 'home laws' [*Heimgesetze*] or 'home councils' [*Heimräte*] actually protect people in total institutions, when it comes down to it" [50].

Like Thomas Galli, whom we cited at the beginning, Heck (himself an insider) found in the emergency both an example of (what he perceived as) the bankruptcy of the system, and a tribune promoting his political agenda, making it visible to, and, most importantly, immediately understandable by, an unprecedentedly large audience, especially of lay people. Quite trivially, the reason for such enhanced understanding was the feeling of shared condition provided, or imposed by the lockdown. "Now we too, have a vague idea", wrote journalist Jörg Kinzig in the *Frankfurter*

Allegemeine Zeitung (April 30, 2020), "what it means to have to stay for a longer period of time in a limited space", with the difference that "the normal citizen still has the privilege to choose with whom to share table and bed" [4].

Yet, the association of total institutions proper with the lockdown as "total institutionalisation" of the whole society has only functioned for a limited time, and only to a limited extent, to further critical attention and civic mobilisation around deinstitutionalisation agendas. Perhaps paradoxically (or, perhaps, not so), the category-turned-catchphrase of "total institution" has fallen victim of its own underdetermination-cum-overemphasis and generalisation in the public sphere. For instance, in her popular private blog "cool aging", journalist and former TV-moderator Rita Werner, already in May 2020 depicted the entire society as a "corona institution": "In a way, this exceptional situation feels like a total institution. In addition, now many want to be Jack Nicholson: for One Flew Over the Cuckoo's Nest" [60].[4] If a blogpost can be forgiven this abuse of rhetoric and metaphorical use of what in the end is a *terminus technicus*, perhaps more disturbing is the fact that a similar confusion has not spared scholars, when addressing the public. Often, the specialist analysis was limited to a blanket suspicion of totalitarianism against institutions and authorities [61], engendering a dissolution of boundaries and a generalisation of the term, through which the idea of the 'total institution' was at times completely decontextualised. On 7 April 2020, for example, the sociologist Rudolf Stichweh diagnosed in the *Frankfurter Allgemeine Zeitung*, a "dramatic break of the corona crisis with the functional differentiation of modern society". Unprecedentedly, the health care system would have been "the whole society as a total institution" [62].

6 Conclusion

Epidemics are the "most political" health phenomenon [63] and the one with the most transformative potential on society. Of the many systemic drawbacks within health care and social systems laid bare by the COVID-19 pandemic, those specific to "total institutions" are the most conspicuous—if only for the exorbitant number of victims. This hecatomb has attracted media attention, and appears to have influenced public attitudes to some extent [19, 64]. If the visibility of "total institutions" during the coronavirus pandemic was literally unprecedented, is a matter for historical comparison to assess. Be it a unique trait of the last pandemic or not, the prominence of "total institutions" may function as a "focusing event" to frame future policy [19].

Our scrutiny of the German public discourse has shown that the use of Goffman's concept has exceeded its original field of application, expanding to encompass, almost metaphorically, the whole of society during the lockdown. At first sight, this

[4] The reference to the Hollywood movie from 1975, shot at the height of international criticism of psychiatry by Milos Forman, was only sensible: the movie is set in a psychiatric hospital, a "total institution", in other words, whose prevailing order is called into question by a new patient, Randle McMurphy/Jack Nicholson.

may seem just a subtler and more erudite reiteration of the trivial, and potentially dangerous mantra "We are all equal in the face of the virus". To be sure, the emphasis on the "anthropological" dimension of vulnerability (as an inherent, common trait of human beings), potentially oblivious of the intersectional complexity of the pandemic event [65], risks overshadowing the "special vulnerability" (as a contingent, extrinsically determined of some groups).[5] In short: it risks concealing the power relations already present in society and "naturalising" social injustice. Nonetheless, as Judith Butler has argued, "common human vulnerability [...] is always articulated differently [and] cannot be properly thought of outside a differentiated field of power and, specifically, the differential operation of norms and recognition" [66, p. 44].

Thus conceived, the (feeling of a) shared "totalising" experience has the potential of shifting the focus away from vulnerable subjects or groups—with all the paternalistic and stigmatising drifts this view can entail—towards the *contexts* of vulnerability. Under this perspective, it would no longer be prisoners, psychiatric patients, people with disabilities, refugees etc., who would be qualified as per se vulnerable, but it would be the contexts of detention, confinement, long term hospitalisation, migration etc., which would be "the cursor for an adapted consideration of the situation experienced by certain people, at a given time, in a given context" [67, p. 242].

The common feeling of a shared experience can be seen as opening a window of opportunity. Building on the work of Judith Butler and Adriana Cavarero on vulnerability, Federica Merenda has recently proposed to consider the perceived shared vulnerability to illness as an "epistemic bonus" [68] for individuals otherwise less inclined to empathically understand the disadvantaged conditions of oppressed groups. Despite the essential difference between forced confinement in psychiatric institutions or imprisonment, and stark restrictions to free movement and sociality, the last pandemic has an unprecedented potential for an "experience-based" and therefore emotional insight into the conditions of people living in long-term isolation. Such an insight has indeed been expressively evoked by some of those individuals: "You (non patients) will now understand us (patients) better, and see for yourselves what it is like to be confined and not meet anyone for days, weeks or even months", was the outburst of some Argentinian patients [64].

This having "seen for ourselves", we argue, has the potential to give a different, more articulated meaning to the definition of the lockdown device as a "social experiment", one not necessarily apocalyptic (as in Agamben's argument), but sustaining a forward-looking perspective. Once the lockdown dispositif has been removed, the possibility remains to take stock of the multilayeredness (cognitive, rational, emotional) of the "shared" experience of "collective totalisation", in order to collectively cognise the structural roots of special vulnerability, thriving in the extant total

[5] See on this distinction and its polical implications the contribution by Henk ten Have in this volume.

institutions. From this perspective, "post-pandemic" is best understood not as a mere chronological concept, but as a possible political programme.

Core Messages

- What Erving Goffman called "total institutions" received sustained media coverage during the coronavirus pandemic, and appears to have influenced public attitudes towards them.
- A certain ambiguity in the use of the expression "total institutions" in the public discourse entails a degree of risk of confusion, as well as of blurring the distinction between the anthropological and special dimensions of vulnerability.
- Nevertheless, the feeling of a shared experience of "totalisation" during the lockdown may function as a "focusing event" to reshape public discourse on "total institutions" and frame future policy.

References

1. Galli T (2018) Sinnvolle und technikbasierte Alternativen zum Vollzug der Freiheitsstrafe in geschlossenen Institutionen. In: Reichenbach MT, Bruns S (eds) Resozialisierung neu denken: Wiedereingliederung straffällig gewordener Menschen als gesamtgesellschaftliche Aufgabe. Lambertus-Verlag, Freiburg, pp 78–93
2. Galli T (2020) Weggesperrt: warum Gefängnisse niemandem nützen. Edition Körber, Hamburg
3. Galli T (2020) Schafft die Gefängnisse ab! Die Zeit, 14 May
4. Kinzig J (2020) Das Unbehagen am Freiheitsentzug. Frankfurter Allgemeine Zeitung, 20 June
5. Bayer F, Müller A, Münster M, Sheld A (2022) Pandemie der Gefängnis-Suizide, Zeit online, 1 November. https://www.zeit.de/politik/deutschland/2022-10/suizide-gefaengnis-selbstmordrate-pandemie. Accessed 31 March 2023
6. Afary F, Al-Kateb L (2020) What is holding back the formation of a global prison abolitionist movement to fight COVID-19 and capitalism? Spectre, 3 April. https://spectrejournal.com/what-is-holding-back-the-formation-of-a-global-prison-abolitionist-movement-to-fight-covid-19-and-capitalism/
7. Bagaric M, Hunter D, Svilar J (2021) Prison abolition. J Crim Law Criminol 111(2):351–406
8. World Health Organization. Regional Office for Europe (2020) Preparedness, prevention and control of Covid-19 in prisons and other places of detention: interim guidance, 15 March 2020. https://apps.who.int/iris/handle/10665/336525. Accessed 31 March 2023
9. Human Right Watch (2020) Announced releases of detainees in the justice system due to covid-19. https://www.hrw.org/sites/default/files/media_2020/05/announced_releases_detainees_covid19.pdf. Accessed 31 March 2023
10. Zeveleva O, Nazif-Munoz JI (2022) COVID-19 and European carcerality: do national prison policies converge when faced with a pandemic? Punishment Soc 24(4):642–666
11. Byrne J, Rapisarda S, Hummer D, Kras KR (2020) An imperfect storm: identifying the root causes of covid-19 outbreaks in the world's largest corrections systems. Victims & Offenders 15:862–909
12. Heard C (2021) Locked in and locked down. Prison life in a pandemic: evidence from ten countries. Project report. ICPR, London, May 2021. https://www.prisonstudies.org/sites/default/files/resources/downloads/locked_in_and_locked_down.pdf
13. Avafia T, Konstantinov B, Esom K, Sanjuan JR, Schleifer R (2020) A rights-based response to COVID-19: lessons learned from HIV and TB epidemics. Health Human Rights J

14. Hanan ME (2020) Incarcerated activism during COVID-19. Ohio State J Criminal Law 18(2):475–513
15. Schneeweis A, Foss KA (2023) "I don't want to die in here": absence and vulnerability in COVID-19 news coverage of prisons. J Commun Inq 47(2):87–206
16. Goffman E (1961) Asylums. Essays on the social situation of mental patients and other inmates. Doubleday, New York
17. Flaker V (2020) Institutions as a lethal weapon: the case of Slovenia in the first wave of the covid-19 epidemic. Revija za socijalna politika 13(16):85–129
18. Lancaster M, Arango E (2021) Health and emotional well-being of urban university students in the era of COVID-19. Traumatology 27(1):107–117
19. Miller EA, Simpson E, Nadash P, Gusmano M (2021) Thrust into the spotlight: COVID-19 focuses media attention on nursing homes. J Gerontol Ser B 76(4):213–218
20. Agamben G et al (2020) Corona virus and philosophers. Eur J Psychoanalisis. https://www.journal-psychoanalysis.eu/discourses/coronavirus-and-philosophers-a-tribune/
21. Agamben G (2020) Wir sollten uns weniger sorgen und mehr denken. Neue Zürcher Zeitung, 7 April
22. Sloterdijk P (2020) Le système occidental va se révéler aussi autoritaire que celui de la Chine. Le Pont, 3 March
23. Lorenzini D (2020) Biopolitics in the time of coronavirus. Crit Inq 47(S2):40–45
24. Quidde L (1900) Arme Leute in Krankenhäusern. Staegmeyr, München
25. Fangerau H, Labisch A (2020) Pest und Corona - Pandemien in Geschichte, Gegenwart und Zukunft. Herder, Freiburg
26. Thießen M (2017) Immunisierte Gesellschaft: Impfen in Deutschland im 19. und 20. Jahrhundert. Vandenhoeck & Ruprecht, Göttingen
27. Foss KA (2020) Constructing the outbreak: epidemics in media and collective memory. University of Massachusetts Press, Boston
28. Ayalon L, Avidor S (2021) 'We have become prisoners of our own age': from a continuing care retirement community to a total institution in the midst of the COVID-19 outbreak. Age Ageing 50(3):664–667
29. Hope J, Schoonhoven L, Griffiths P, Gould L, Bridges J (2022) 'I'll put up with things for a long time before I need to call anybody': face work, the total institution and the perpetuation of care inequalities. Sociol Health Illn 44(2):469–487
30. McGahey B (2021) 'It's like waking up in the library': how an international student dorm in Copenhagen became a closed circuit during COVID-19. Anthropol Action 28(1):63–66
31. Pepper A, Voigt K (2021) Covid-19 and the future of zoos. Les ateliers de l'éthique 16(1):68–87
32. Marazia C, Fangerau H et al (2020) Visions of another world. Franco Basaglia and the German Reform. In: Burns T, Foot J (eds) Basaglia's international legacy: from asylum to community. Oxford University Press, New York, pp 227–247
33. Gijswijt-Hofstra M, Oosterhuis H, Vijselaar J (eds) (2005) Psychiatric cultures compared: psychiatry and mental health care in the twentieth century: comparisons and approaches. Amsterdam University Press, Amsterdam
34. Hinds L (2011) Crime control in Western countries, 1970 to 2000. In: Pratt J, Brown D, Brown M, Hallsworth S, Morrison W (eds) The new punitiveness. Routlege, New York, pp 47–65
35. Dollinger B, Schmidt H (2015) Zur Aktualität von Goffmans Konzept "totaler Institutionen" – Empirische Befunde zur gegenwärtigen Situation des "Unterlebens" in Gefängnissen. In: Schweder M (ed) Handbuch Jugendstrafvollzug – Handlungsfelder und Konzepte. Beltz Juventa, Weinheim Basel, pp 245–259
36. Schulz-Nieswandt F, Köster U, Mann C (2021) Lehren aus der Corona-Krise. Modernisierung des Wächterstaates im SGB XI. Nomos, Baden-Baden
37. Schülein JA (2007) "Asyle" - Über Goffmans Analyse und Kritik sozialer Ausgrenzung und Kontrolle. Österreichische Zeitschrift für Soziologie 32:32–52
38. Bretschneider F (2008) Die Geschichtslosigkeit der "Totalen Institutionen". Kommentar zu Erving Goffmans Studie "Asyle" aus geschichtswissenschaftlicher Perspektive. Wiener Zeitschrift zur Geschichte der Neuzeit 1(08):135–143

39. Kitsotaki D, Long V, Smith M (eds) (2016) Deinstitutionalisation and after: post-war psychiatry in the Western world. Palgrave Macmillan, London
40. Ellis R (2021) Prisons as porous institutions. Theory Soc 50(2):175–199
41. Scherrer G, Vögeli D (2022) Drinnen der Wahnsinn, draussen die Normalen – das war einmal. Ein Tag in der grössten psychiatrischen Klinik der Schweiz. Neue Zürcher Zeitung, 28 Dec
42. Heidegger M, Dietrich-Daum E (2008) Die k. k. Provinzial-Irrenanstalt Hall in Tirol im Vormärz – eine Totale Institution? Wiener Zeitschrift zur Geschichte der Neuzeit 8(1):68–85
43. Romania V (2020) Interactional anomie? Imaging social distance after COVID-19: a Goffmanian perspective. Sociologica 14(1):51–66
44. Redaktion (2020) Die Gesellschaft im Krankenbett. Frankfurter Allgemeine Zeitung, December 30
45. Wagle F (2020) Pflegeheime im Lockdown – Orte der Einsamkeit? Bayer Aztebl 7–8:332–335
46. Heinzelmann M (2004) Das Altenheim – immer noch eine "Totale Institution"? Eine Untersuchung des Binnenlebens zweier Altenheime. Univ. Diss. Sozialwissenschaft, Göttingen
47. Rödler P (2020) Totale Institution: Die Renaissance der geschlossenen Unterbringung im Zuge der Corona Pandemie? Behindertenpädagogik 59(4):345–358
48. Köbsell S, Achtelik K (2020) Institutionen werden wieder total. Der Freitag, 19 May
49. Feest J (2020) Corona und Knast – ein Zwischenbericht Corona-Status 2021 – Wie die Regierung hoch vulnerable Gruppen opfert. Neue Kriminalpolitik 32:113–122
50. Heck K (2020) Schutzhaft in Pflegeheimen. Telepolis, April 23. https://www.heise.de/tp/features/Schutzhaft-in-Pflegeheimen-4707806.html?seite=all. Accessed 31 March 2023
51. Valin F (2022) Ein Haus voller Wände. Verbrecher Verlag, Berlin
52. Wohlrab-Sahr M (2021) Generalisierter Anderer oder gespürter Nächster. Paradoxien der Sozialität in der Coronakrise. In: Lenz S, Hasenfratz M (eds) Capitalism unbound: Ökonomie, Ökologie, Kultur. Campus: Frankfurt/New York, pp 265–283
53. Trescher H, Nothbaum P (2022) Institutionalisierte Lebenslagen von Menschen mit geistiger Behinderung und Perspektiven pädagogischen Handelns während der COVID-19-Pandemie. Behindertenpädagogik 61(2):137–157
54. Maskos R (2021) Ignoriert: Behinderte Menschen in der Pandemie. Blätter für Deutsche und Internationale Politik 4 April. https://www.blaetter.de/ausgabe/2021/april/ignoriert-behinderte-menschen-in-der-pandemie
55. Adorjan K et al (2021) Zwangsmaßnahmen und die SARS-CoV-2-Pandemie. 92(5):501–506
56. European Committee for the Prevention of Torture and Inhuman or Degrading Treatment or Punishment (CPT) (2020) Statement of principles relating to the treatment of persons deprived of their liberty in the context of the coronavirus disease (Covid-19) pandemic, 20 March 2020. https://rm.coe.int/16809cfa4b
57. Bundesarbeitsgemeinschaft für Straffälligenhilfe e.V. (2021) Priorisierung von Gefangenen bei Corona-Impfung, 1 Feb. https://www.bag-s.de/aktuelles/aktuelles0/priorisierung-von-gefangenen-bei-der-corona-impfung. Accessed 31 March 2023
58. Röttger-Liepmann B, Hopfmüller E (eds) (2002) Initiative zur Einrichtung einer "Enquête der Heime": Dokumentation einer Tagung am 21.03.2002. Institut für Pflegewissenschaft an der Universität Bielefeld (IPW), Bielefeld
59. Bühring P (2001) Versorgung in Heimen: Grundrechte bedroht. Deutsches Ärzteblatt 98(31–32):2004–2005
60. Werner R (2020) Totale Institution – willkommen in der Coronanstalt? 12 May, Cool Aging. https://cool-aging.de/totale-institution/. Accessed 31 March 2023
61. Schrappe M (2022) Verlorene Jahrzehnte – Zehn Thesen zur Aufarbeitung der Corona-Krise. Cicero Online. Magazin für politische Kultur, Dec 1
62. Stichweh R (2020) An diesem Imperativ kann die Politik scheitern. Frankfurter Allgemeine Zeitung, 7 April
63. Vasold A (2015) Grippe, Pest und Cholera. Eine Geschichte der Seuchen in Europe. Steiner, Stuttgart

64. Fernandez MA, Ares Lavalle G, Murlender L, Agrest M, Ardila-Gómez SE (2023) 'Now you will understand what it is like to be confined': did COVID-19 lockdowns affect perceptions about long-term psychiatric hospitalizations? A report from Argentina. Int J Soc Psychiatry. First online. https://doi.org/10.1177/00207640221143291
65. Crenshaw K (1989) Demarginalizing the intersection of race and sex: a black feminist critique of antidiscrimination doctrine, feminist theory and antiracist politics. Univ Chic Leg Forum 1:139–167
66. Butler J (2004) Precarious life: the powers of mourning and violence. Verso, London/New York
67. Burgorgue-Larsen L (2014) La vulnérabilité saisie par la philosophie, la sociologie et le droit de la nécessité d'un dialogue interdisciplinaire. In: Burgorgue-Larsen L (ed) La vulnérabilité saisie par les juges en Europe. Pedone, Paris, pp 237–243
68. Merenda F (2022) Imparare dalla vulnerabilità. Le passioni come bonus epistemico e motivazioni all'agire politico nel contesto pandemico. Biblioteca della libertà 235:5–38

Chantal Marazia is assistant professor in the History Philosophy and Ethics of Medicine at the University of Düsseldorf (Germany). Her research focuses are the history of psychiatry and neurology, and the history of human rights. She is editor of the psychopathological writings of Karl Jaspers.

Nils Löffelbein is a historian and works as a scientific researcher at the Department of the History, Philosophy and Ethics of Medicine at Heinrich-Heine-University in Düsseldorf. His work focuses on the history of psychiatry and disability care in Germany after 1945, material culture research, the history of the GDR and the history of the war victims of the First World War.

Heiner Fangerau is chair and director of the Department of the History, Philosophy and Ethics of Medicine, Heinrich-Heine-University Duesseldorf. He held chairs in the history, philosophy and ethics of medicine in Ulm (2008–2014) and Cologne (2014–2015). He has been president of the European Association for the History of Medicine and Health and the German Society for the History of Science. He is a member of the editorial board of the Journal of the History of Medicine and the Allied Sciences and the European Journal of the History of Medicine and Health. In 2017 he was elected as a member of the German National Academy of Sciences Leopoldina. His main research fields are history and ethics of psychiatry and neurology, history of the life sciences, child and adolescence psychiatry, medical diagnostics and historical network analyses.

Part III
The Future of Vulnerability

Chapter 12
Vulnerable to Ourselves, or the Radicalized Disenchantment of Being

Kamran Baradaran

> *I think we're in real trouble. I don't know how this started or why,*
> *but I know it's here and we'd be crazy to ignore it.*
> *The bird war, the bird attack, Plague - call it what you like.*
> *They're amassing out there someplace and they'll be back.*
> *You can count on it*
> *Alfred Hitchcock, The Birds, 1963*

Abstract This article presents the concept of vulnerability as a core stone for politics. To do so, it engages with Slavoj Žižek's contemporary reading of Hegel and Lacan. The goal of doing so is to demonstrate how inherent vulnerability interprets the nature of subjectivity, intersubjective relations and the relationship between the subject and the world it inhabits. This article also tries to show that the only way to break the vicious cycle of vulnerability is to focus on a new definition of political act to abolish and deconstruct the capitalist relations and the very circumstances that make us vulnerable.

Keywords Slavoj Žižek · G.W.F. Hegel · Judith Butler · Capitalism · COVID-19 pandemic · Jean-Luc Nancy

K. Baradaran (✉)
Department of Teaching English as a Foreign Language, Faculty of Foreign Languages, North Tehran Branch, Islamic Azad University, Tehran, Iran
e-mail: kamranbaradaranrazaz@gmail.com

S. Achella and C. Marazia (eds.), *Vulnerabilities*, Integrated Science 18,
https://doi.org/10.1007/978-3-031-39378-5_12

1 Introduction

The Cocoanuts (1929) by Marx Brothers contains a wonderful scene that I think best illustrates our current impasse; Mr. Hammer (Groucho) is discussing with Chico about a map, and Hammer mentions the presence of a viaduct between the mainland and a peninsula:

Hammer: …Now, here is a little peninsula, and, eh, here is a viaduct leading over to the mainland.

Chico: Why a duck?

Hammer: I'm alright, how are you? I say, here is a little peninsula, and here is a viaduct leading over to the mainland.

Chico: Alright, why a duck?

Hammer: (pause) I'm not playing "Ask Me Another," I say that's a viaduct.

Chico: Alright! Why a duck? Why that…why a duck? Why a no chicken?

Hammer: Well, I don't know why a no chicken; I'm a stranger here myself. All I know is that it's a viaduct. You try to cross over there a chicken and you'll find out why a duck.

Chico: When I go someplace I just…

Hammer: (interrupts) It's...It's deep water, that's why a duck. It's deep water.

Chico: That's why a duck…

Hammer: Look...look, suppose you were out horseback riding and you came to that stream and you wanted to ford over...You couldn't make it, it's too deep!

Chico: Well, why do you want with a Ford if you gotta horse?

Hammer: Well, I'm sorry the matter ever came up. All I know is that it's a viaduct.

Chico: Now look, alright, I catch ona why a horse, why a chicken, why a this, why a that... I no catch ona why a duck.

Hammer: I was only fooling...I was only fooling. They're gonna build a tunnel there in the morning.

Isn't this the same predicament we are in today? In recent years, scientists and ecologists have repeatedly warned us of the dangers of the current way of life in capitalism and how ignoring them could result in a full out disaster, whether environmental or economic and health crises. Many scientists have repeatedly warned us of the dangers of a full-scale catastrophe that will be economically, ecologically and politically devastating. However, we reacted to these warnings with the structure of fetishist disavowal: catastrophes, from Fukushima Dai-ichi nuclear disaster to the current case of Covid-19 pandemic, struck and we still wonder "why a duck?".

Writing about disaster and crisis is not a difficult task these days. Since the COVID-19 pandemic struck mankind, much has been said and written about the concept of disaster. The same goes for the concept of vulnerability; from Plato to Hegel, Levinas, and Foucault, vulnerability has inhabited the philosophical landscape. Often associated with violence, finitude, or mortality, this concept is often used to describe the feeling of being susceptible to injury or feeling threatened. My intention here is to present a comprehensive picture of the nature and ideology of vulnerability, emphasizing the prior status of the vulnerable subject in relation to contemporary politics and capitalism. In the following pages, I will argue that vulnerability is not only an integral part of the process of subjectification but also the best starting point for radical politics and action, a kind of launching pad or starting station for a long and critical journey.

2 Fatal Strategies

During the height of the pandemic, one couldn't help but to notice that even politicians had more or less abandoned the promises of the sweet days after the end of the health predicament. All the leading figures of the so-called *Realpolitiks* were warning of more cataclysmic distress in the future, from unemployment and the financial crash to a dangerous and extensive psychological breakdown. It seems that even the idea of good days to come has disappeared from the political sphere, and the state of constant vulnerability has become the dominant worldview among politicians and experts.

The first problem when a catastrophe occurs, when our vulnerability manifests itself, is the ability to live in a world where that end is a necessity: whether it is the end of humanity, the end of an age, or the end of history. The question of the end has a kind of internal contradiction in itself. Beginning-end, like other dichotomies (subject-object, cause-effect), leads to a form of consolation. Do not we all take comfort in knowing that one day everything will all come to an end, when the linearity of time will reach its limit? As Haruki Murakami once said, "everyone, deep in their hearts, is waiting for the end of the world to come" [1, p. 120]. A sense of responsibility, however, can result from approaching the end, a sense that can prevent the inevitable.

However, today, even the possibility of end and disintegration has been taken away from us. History is no longer even coming to an end but is reversing and destroying all the events that were once considered part of the political and ideological subconscious of the unfortunate people of this pitiless era. It is clear that neither pervasive social movements nor bleak environmental or nuclear catastrophes, nor even the current health crisis, can revive our hope or possibility of an event. This shows that "we are engaged in a gigantic process of *revisionism*—not an ideological revisionism but a revisionism of history itself" [2, p. 32].

The current COVID-19 epidemic is a clear example of this desperate deadlock. First, one must note that COVID-19 is pure materialism, a process devoid of meaning, something that simply happens, but it does so under certain economic conditions. Second, any attempt to justify the COVID-19 pandemic from an ontological point of view leads to an impasse; the health emergency we are experiencing has no logical basis except the mechanics of liberal capitalism, and any attempt to link it to an ontological narrative ("this pandemic is a sign of the decline of humanity at the global level" and so on) will lead nowhere. COVID-19 is as senseless as its main cause, i.e., capitalism, and the only way to face it is to understand this senselessness as its inherent condition.

At least at first, it was possible to hope that around the world, in spite of such an abysmal situation, there would be an increase in solidarity and cooperation. Many people remember the ridiculous images of people singing from balconies or police officers dancing and cheering in neighborhoods trying to lift the spirits of those trapped during long quarantines. The Guardian, however, revealed the meaninglessness of such images:

> A few days into Italy's lockdown, people across the country sang and played music from their balconies as they came together to say "Everything will be alright" (*Andrà tutto bene*). Three weeks on, the singing has stopped and social unrest is mounting as a significant part of the population, especially in the poorer south, realize that everything is not all right. [3]

In his *A Contribution to the Critique of Hegel's Philosophy of Right*, Karl Marx talks about the decline of the political powers of Germany as the ironic repetition of the tragic collapse of the French ancient régime:

> The present German regime, on the other hand—an anachronism, a flagrant contradiction of universally accepted axioms, the futility of the *ancient régime* displayed for all the world to see—only imagines that it still believes in itself and asks the world to share in its fantasy. If it believed in its own nature, would it try to hide that nature under the appearance of an alien nature and seek its salvation in hypocrisy and sophism? The modern *ancient régime* is merely the clown of a world order whose real heroes are dead. History is thorough and passes through many stages while bearing an ancient form to its grave. The last phase of a world-historical form is its comedy. The Greek gods, who already died once of their wounds in Aeschylus's tragedy *Prometheus Bound*, were forced to die a second death—this time a comic one—in Lucian's Dialogues. Why does history take this course? Therefore, mankind may part happily with its past. We lay claim to this *happy* historical destiny for the political powers of Germany. [4, pp. 247–248]

The formula of a regime that "only imagines that it believes in itself" shows the current situation in the world after the COVID-19 pandemic. Today, governments and their supporters only imagine that they believe in their own sufficiency in crisis management, but in practice, it can be said that everyone has come to the conclusion that humanity is dealing with a kind of global farce in the field of dealing with crises.

This of course reminds one of the famous scene in *Onibaba* by Kaneto Shindō: The older woman discovers that, after getting wet in the rain, her *Hannya* mask that she used to scare others with is impossible to remove. She reveals her scheme to her daughter-in-law and pleads for her to help take off the mask. After failing to pull it off, the young woman breaks off the mask with a hammer. Under the mask, the older woman's face is now disfigured. Is not this the perfect manifestation of the predicament we are facing today? Finally, the mask has been broken, and what we have is the mutilated face of a world we once considered perfect and flawless (at least for a certain group of people).

In his *Histoire de la merde*, French psychoanalyst Dominique Laporte shows how the development of sanitation techniques in Western Europe affected the formation of modern notions of individuality [5]. One might wonder if we can consider the new forms of sanitation techniques amidst the COVID-19 pandemic as a means to formulate a new form of individuality. Would this new form of the individual be the "subject"(s) for the coming community? This new form of individuality would not be based on a disciplinary apparatus (as developed by Michel Foucault) but could easily lead to new forms of apartheid and oppression. The future that awaits most of us is far worse than any form of the so-called *Society of the Spectacle*. One might be hopeful of a wave of sympathy and compassion after a disaster, but the true courage lies in understanding hopelessness in embracing the ultimate negativity that comes from total destruction: There is an infinite amount of solidarity under capitalism … but not for us!

3 The Lessons of Catastrophic Exposure

As Jean-Luc Nancy has shown [6, p. 128], that which is shattered and that which shatters must first be held in experience without seeking to master it in order to understand it. At the same time, endeavors to grasp our current predicament should not be characterized by a retreat of philosophy to the ethically fraught science of *calypsology*, namely, "the science of the concealment of a thing such that it is not open for other contenders; such a discipline would be confined to the immurement of its object away even from itself" [6].

In regard to the issue of "vulnerabilities", an important question inevitably arises: To what is vulnerability due? Therefore, one can say that the vulnerability paradigm now includes a number of characteristics that can relate to very different levels (natural, political, cultural, etc.). Thus, a "vulnerable subject" can be defined against a deadly disease, against a terrorist whose murderous plans must be thwarted, or even against nature, which will soon lead us out of its cycle, or even against an apparatus that threatens our lives.

Does not the current situation, what has happened to humanity over the past decade, signify that we have all become as vulnerable as Fyodor Dostoevsky's the *underground man*? Events such as the Fukushima nuclear disaster and the recent example of COVID-19 have shown that, with a few exceptions, most people are nothing more than the poor and worthless masses trapped in the intricate mechanism of the ruling order. We are counted in the middle of this commotion, but we are not *accounted for*. Reality can appear only in such moments, in total alienation, in an underground world without any clear vision; it is in such a desperate situation that there is the will to act. The dream of the burning child in Sigmund Freud's *Die Traumdeutung* portrays our situation perfectly:

> A father had been watching beside his child's sick-bed for days and nights on end. After the child had died, he went into the next room to lie down, but left the door open so that he could see from his bedroom into the room in which his child's body was laid out, with tall candles standing round it. An old man had been engaged to keep watch over it, and sat beside the body murmuring prayers. After a few hours' sleep, the father had a dream *that his child was standing beside his bed, caught him by the arm and whispered to him reproachfully: 'Father, don't you see I'm burning?'* He woke up, noticed a bright glare of light from the next room, hurried into it and found the old watchman had dropped off to sleep and that the wrappings and one of the arms of his beloved child's dead body had been burned by a lighted candle that had fallen on them. [7, p. 353]

First, the subject constructs a dream, a story that enables him to prolong his sleep, to avoid awakening into reality. However, the thing he encounters in the dream, the reality of his desire, the Lacanian Real—the reality of the child's reproach to his father—is more terrifying than so-called external reality itself: to escape the Real of his desire, which announces itself in the terrifying dream [8, p. 45]. This is exactly how we operate as narrators of a crisis. When we talk about the happy days that come after the end of a catastrophe, we act according to the same logic. Essentially, just like the father, we awaken from a ruined world so we can carry on 'dreaming', in the sense that by returning to 'external reality', we can escape from something even more horrifying.

Bohemian Landscape with Mount Milleschauer, Caspar David Friedrich (1808)

Oxana Timofeeva once said that catastrophe defines the borders of a collective and the true sense of what we call history [9]. There is more truth in this statement than meets the eye. After all, the linear process of time is broken only through the evental turns that create a sense of change and entry into a new era, i.e., "fatal strategies of time" that shape our ways of thinking about history and its imaginary end. As Timofeeva puts it brilliantly, "the very word *Apokalypsis*, from the *Koine* Greek, means 'unveiling' or 'revelation.' It unveils and reveals the truth about a certain reality. As far as it unveils (i.e., unveils what is), etymologically, the apocalypse is always now" [9]. Disaster is always associated with the issue of "salvation"; the apocalypse is finally here, and the forces of good and evil will have their final confrontation, an event that will determine the fate and end of the world. However, what if we take a more radical angle? A truly materialistic action is not rejecting the idea of the end of time and disaster but secularizing it, emptying it of the theological aspect of salvation and accepting this end as an event that does not lead to any liberation or catharsis, and its mere existence indicates the impossibility of such a positive negativity.

According to Jean-Pierre Dupuy, our failure to prevent looming climate catastrophe results from a faulty metaphysics of time. Dupuy believes that "we have irreversibly entered into an era that its ultimate prospect is the self-destruction of the human race" [10]. Accordingly, because we think there are many possible futures, some of which bypass the catastrophe, we do not believe it is absolutely urgent to take drastic action now. Dupuy argues that to confront the disaster, we should first perceive it as our unavoidable fate, accepting that, at the level of possibilities, our future is doomed. Then, with our affirmation of the forthcoming catastrophe, we should mobilize ourselves to perform the act that will change destiny itself and thereby insert a new possibility into the past.

COVID-19 has triggered an unprecedented humanitarian emergency and put us in an uncertain global environment. This uncertainty has a clear reason. In our lifetime, we have never faced a pandemic of this magnitude. Individuals living today have no previous experience with a major pandemic, as the last one occurred in 1918–1922, exactly 100 years ago. Here, we have to make a kind of universalization: do not all existing structures, from parents to school, university, work environment, marriage, etc., lead to some kind of "vulnerability"? In all these cases, are not we dealing with a process that subjugated the subject to an authority in the very process of establishing it as free and autonomous? There is always some libidinal dirt staining the ideal figure of a situation. Therefore, in a more accurate generalization, one might even say that what is ultimately traumatic in the network of vulnerability is the concept of *Neighbor* in the strict sense of the word, the abyss of its desire and its obscene enjoyment. Vulnerability always raises the issue of the other, and the ultimate goal of all systems of dominance is to neutralize the "toxic" dimension of the other and reduce it to a fellow man, an attempt to violently remove the *object petit a* that forms the consistency of the subject.

For Slavoj Žižek, subjectivity itself can be thought of as a source of contingency or instability, which means that intersubjective relations, all the way up to political communities, could also be characterized as contingent. This is what he calls "negativity" of the subject in which the subject as the source of negativity becomes reimagined as relationships as the embodiment of vulnerability or, in other words, a form of ontological and subjective vulnerability. As Žižek puts it perfectly, "the subject is the power of negativity, of introducing a gap/cut into the given-immediate substantial unity" [11, p. 106]. Accordingly, Spirit is the result of its own activity, which is the transcending of what is immediately there, by negating it and returning into itself. The important point here is the emphasis that is placed on the "negativity" of the subject. This negativity implies an existential inability, or vulnerability, as a result of which the individual can never achieve a completely fixed or stable identity/subjectivity, and by extension, a society at large can never achieve a fixed or stable ideological framework to bind subjects to a social framework. This substantive negativity gives birth to ontological and subjective vulnerability. This "negativity" should be understood as not merely a characteristic of an individual subject but rather as the result of the dynamics of intersubjectivity itself, which means that any human relationship is a constantly evolving entity resulting from this vulnerability. Here, the void functions as the principle of subjectivity, or as Hegel remarks, "void as the principle of motion":

> The disparity which exists in consciousness between the 'I' and the substance which is its object is the distinction between them, the negative in general. This can be regarded as the defect of both, though it is their soul, or that which moves them. That is why some of the ancients conceived the void as the principle of motion, for they rightly saw the moving principle as the negative, though they did not as yet grasp that the negative is the self. [12, p. 37]

The idea of "the void as the principle of motion" shows that the essential condition for moving forwards is not emphasizing positivity but pervasive and inherent negativity; subjectivity as the principle of doubt, an empty position that negates all determinations.

In *Vulnerability: New Ethics and Feminist Philosophy,* Catriona Mackenzie, Wendy Rogers and Susan Dodds attempt to tackle the problem by distinguishing between three types of vulnerabilities: inherent vulnerability, which is "intrinsic to the human condition;" situational vulnerability, which is "context specific"; and pathogenic vulnerability, which stems from abuse, injustice or oppression [13]. An important point should be noted here; the mentioned distinction should place its starting point on inherent vulnerability and consider it as the basis of any susceptibility. In other words, acceptance of fundamental vulnerability is the key to overcoming the predicament we are all in. This is the foundation of any ethical activity par excellence. The link between Hegel's concept of subject as a void and Lacan's concept of the Real is based on the idea that there is a "gap" in human subjectivity that can never be permanently filled by a particular system of meaning, which means that there is no external point of reference from which we could perceive the relativity of our own "merely subjective" standpoint [11, p. 393].

Untitled, by Bahman Mohasses. *Source* Wikiart

According to Judith Butler, we cannot understand bodily vulnerability outside of the conception of relations [14]. This kind of "improvisation within a scene of constraint" [15, p. 1] implies that the self is not a mere individualist agent in the world but instead a play of unpredictability, possibility, and network of relations that acknowledges freedom from social or universalized norms. Butler's thesis has many positive points and well depicts the complex state of vulnerability of the subject. However, it should be given another twist and made more radical. In other words, Lacanian *symbolic castration* is a necessary and fundamental condition of vulnerable subjectivity; instead of an agent who will give structure to our chaotic social lives, we must navigate our vulnerabilities and the very ways in which we frame and structure our understanding of the world through rigorous, hermeneutic questioning. A complete self-explanation in a symbolic narrative is basically not possible a priori because the position of the subject depends on the connection it has with the other, and in this way, the nature of the subject is always vulnerable. Being caught in the complex network of relations is the condition of the subject's autonomy. As Stefania Achella demonstrates brilliantly, confrontation with this Hegelian "night of self-preservation" is the condition of possibility for the living being to become human. Consequently, the process of subjectification must come to terms with this condition of darkness and unconsciousness [16].

Due to the pandemic, empty streets, parks, hotels, airports, desolated shopping malls, and somewhat vacant cities are new examples of ruins in our time. However, unlike in the classical period, in the modern era, these ruins are no longer reminiscent of the lost glory of ancient Greece or Rome but show the destruction that is at the heart of the idea of progress in the capitalist world. Is not the disappearance of the aura of the inner world of capital a way to politicize this situation and try not only to overcome it but also to focus on similar catastrophes of the future? This "political" act comes to existence only when, as Slavoj Žižek mentions in his *Enjoy Your Symptoms*, the subject suspends the network of symbolic fictions that serve as support for his daily life and again confronts the radical negativity upon which they are founded [17].

Our vulnerable situation is reminiscent of Buster Keaton's famous falls, which represent unbridled dislocation. Anyone with a passing interest in comedies can immediately perceive that Keaton has a distinct position among his contemporaries. Keaton, with his soulless and anemic face, was an expert at falling and keeping his equilibrium. In Keaton films, one thing that always stands out is that even when everything has died and gone and there is nothing left to laugh about, Keaton's position remains as it was: falling. In his films, Keaton usually has no past or future. As opposed to Lloyd or Chaplin, he usually does not belong to a definite class or have particular cultural characteristics. Keaton is not even lucky enough to pick up a red flag to be mistaken for the leader of a workers' demonstration such as Charlie Chaplin in *Modern Times* (1936). For Keaton, the world is nothing but a free fall and unbridled downhill. There is no sign of a sense of victimhood or oppression on Keaton's face; it was no coincidence that Beckett was so fond of him.

Today, we should face our vulnerability in a similar manner to Keaton: With no foundation to cling to, we may be left with no escape route but a downhill fall. Keaton's specter still roams Fukushima's ruins, Chernobyl's decaying structures, and the world's most uninhabited promenades today like a ghost, a symbol of our fundamental vulnerability.

In contrast to the subject's entanglement in a network of relations, Žižek's view provides a better position: the subject can suspend the determination and uniformity of the other. In this context, vulnerability has a completely different meaning; separation of the vulnerable subject (from the homogeneous symbolic world) and the resulting limitations make the subject able to occupy the place of absence in the symbolic other. Therefore, this limitation not only limits our humanity but is also its affirmative condition. Recognizing the other does not mean recognizing in a specific and defined form but rather recognizing the other in the abyss of its ambiguity and anonymity.

A new understanding of substance and subject is a step towards a new understanding of relationships with one another. Today, dealing with the concept of "other" in the framework of "vulnerability" is basically defined in the form of concepts such as charity and humanitarian affairs, that is, the help of human rights institutions and NGOs to bring the vulnerable out of its fragile situation. As a consequence, moralist convictions such as "the brotherhood of man", "universal peace", and "justice for all" are often used in a hollow manner as empty catchphrases to evade the responsibility of ethical reflection. As far as the ideology of human rights is concerned, one should point out the duplicity involved in campaigns that disguise their motivation behind a veil of pseudo-love for the other. Hence, charity is merely a humanitarian mask, covering up economic exploitation beneath.

Elevating the other as human beings like us basically leads nowhere. There is no direct and immediate relationship between the vulnerable and others; in contrast, this relationship always requires the existence of an impersonal symbolic order. This is where the politics of vulnerability becomes important: without this third factor, the other becomes something monstrous. Because the third factor has a decisive value in this relationship, it should be criticized and inspected. Otherwise, we will end up with nothing but a boring world where subjects have become soulless soldiers of regulated communication. To accept the neighbor, one should accept the concept of the "real" Other. This means that the concept of "reality" is only accurate when the "real" is known and close; otherwise, it is just an illusion since the truth behind the "real" is veiled. Slavoj Žižek links the concept of "reality" and the "real" by the idea of *jouissance*, the experience when reality keeps you away from the real, creating pleasure. Jacques Derrida also once argued that "a finite being could not possibly be present in act to too great a number. There is no belonging or friendly community that is present, and first *present* to *itself, in act*, without election and without selection" [18, p. 21]. It is easy to love an idealized figure of a poor, vulnerable neighbor, an African or Indian starving to death; that is, it is easy to love my neighbor if he lives far enough away from me, if there is a reasonable distance between us. The problem arises at the moment when the neighbor gets too close, when we start to feel his/her stifling contiguity, i.e., when the neighbor exposes him/herself to us too much and love suddenly turns into hatred. Aversion to one's own vulnerability and lack might be mirrored in another, causing one to avoid encountering the unique and concrete experiences of the neighbor. Because of this avoidance, humanitarian causes are popular due to their paradoxical nature, as they are able to be loved from a distance without getting directly involved. Is not this the ultimate paradox of what is dubbed "postcolonial" studies, namely, that they reduce the ultimate Neighbor to an idealized poor figure that can only be determined with minimal tools without dealing with the real causes and effects of colonialism, namely, today's global capitalism?

In a scene from the movie *Casablanca*, Victor Laszlo, a renowned, fugitive Czech resistance fighter, is talking with Captain Renault after arriving in Casablanca. "Did you have a good night's rest?" Renault asks. "I slept very well!" Laszlo replies. In addition, there comes the wonderful remark of Captain Renault: "*That's strange; nobody's supposed to sleep well in Casablanca.*" The logic behind this dialogue is, of course, very simple: what Renault is trying to express in metaphorical language is that the Nazi regime has created so much suffocation that no one can live a comfortable life. That is, "sleeping badly" is the unfortunate result of living in this situation. Today, the only people who can easily go to bed at night, without fear and haunting nightmares, are those who are probably hidden in their shelters in New Zealand, while ordinary people have to face many dangers and fears every day. Thus, our basic dogma should be that today everyone has to act as if they are not supposed to sleep well. This means that we have to accept that escaping from our everyday vulnerable situation is not possible, at least until the state of the world follows such a miserable trend. The ruins of our fragile state must become a part of everyday life, and we must not be fooled by attempts to integrate the exposed into silly humanitarian activities designed to divert attention from the root cause of this state. No one is meant to sleep well, especially since we all suffer from an innate vulnerability.

As David Graeber once said, "the crisis we just experienced *was* waking from a dream, a confrontation with the actual reality of human life, which is that we are a collection of fragile beings taking care of one another and that those who do the lion's share of this care work that keeps us alive are overtaxed, underpaid, and daily humiliated and that a very large proportion of the population don't do anything at all but spin fantasies, extract rents, and generally get in the way of those who are making, fixing, moving, and transporting things, or tending to the needs of other living beings. It is imperative that we not slip back into a reality where all this makes some sort of inexplicable sense, the way senseless things so often do in dreams" [19]. In short, nobody's supposed to sleep well in the middle of this mess, because we simply can't slip back into a meaningless dream where the way our society is organized is seen as inevitable!

There are obvious signs of the nightmare that await us in the future. On 27 October 2020, Guardian reported that scientists have found evidence that frozen methane deposits in the Arctic Ocean have started to be released over a large area of the continental slope off the East Siberian coast. According to this report, high levels of the potent greenhouse gas have been detected down to a depth of 350 m in the Laptev Sea near Russia, prompting concern among researchers that the discovery could have "serious climate consequences" [20]. What makes the story even scarcer is the fact that methane has a warming effect 80 times stronger than carbon dioxide over 20 years. According to Guardian, since the Arctic is now warming twice as fast as the global average, there is a fear that methane will be released into the atmosphere, although the timing has been a matter of considerable uncertainty in climate computer models. This means that with the current trend of global warming, not only man-made structures but also nature itself can become part of the problem as a result of climate change. Of course, it can be argued that all this is nothing but a pessimistic speculation, but in the current situation, such a dark view is needed. As Srećko Horvat once elaborated perfectly, "we need to be mad prophets who might turn out wrong. We need to shock the people with the dystopian facts. No sea fish, only plastics, no air" [21].

4 Politics of Vulnerability

How does the idea of a fundamental human susceptibility enable us to account for socially produced or configured forms of vulnerability? What does this radical detachment mean for our future? Wagner includes a wonderful phrase at the end of his *Parsifal* opera: "The wound can be healed only by the spear or sword that inflicted it". This means that the very disintegration of traditional forms of life opens up a space for liberation. The same goes for all of the impractical and idiotic nostalgia for the conditions prior to the pandemic. While this widespread disease was a devastating blow, return to the idealized past is not possible. Instead, we have to see this deprivation (of the days of the past) as a unique chance of freedom and try to embrace the authenticity of the ruins we see unfolding before us [22].

The COVID-19 pandemic, along with other events of the twenty-first century, such as the financial collapse of 2008 and the terrorist attacks of September 11, once again rejected the absurd idea that in the current era, the apocalypse and catastrophe have no place in the course of history; therefore, valid changes are not necessary. As Sacha Ghandeharian and Maggie FitzGerald recently elaborated, vulnerability must be understood as a fundamental political concept that merits ongoing attention in our political systems [23]. Postmodern society is a society of uncertainty in which the ongoing transformations have led to an erosion of the certainties of modern society. One can only agree with Robert Castel, who argues that current uncertainty is the effect of the gap between a socially constructed expectation of protections and a society's actual ability to make them work [24].

Our precarious predicament has given new life to one of the Left's oldest concepts: antagonism. One should note the undeniable fact that in the face of viral infections, it was easy for many of us, with the means to self-isolate, to accept lockdowns and quarantines. However, what about those who were not able to do so? What about the "essential workers" and others who were forced to keep working in the current situation just to stay alive and to keep us alive as well? Here, there is no better tool than that of class antagonism. To reintroduce this classical category, we need a new radical form of political action, an intervention that changes the very framework that determines how things work. This intervention enables us to go beyond the "normal order of things". As Rancière puts it, "it is this anomaly that is expressed in the nature of political subjects who are not social groups but rather forms of inscription of the (ac)count of the unaccounted-for" [25, p. 35].

As far as the mechanism based on liberalism is concerned, was not the global economic crisis of 2008 a sign that we are all inherently vulnerable? Was not the catastrophe of CDOs in the American economy itself clear evidence that not only do we face uncertainty and vulnerability in our everyday lives, but even the future plans that we have are inherently doubtable? We may lose everything, and each of us might end up becoming worthless units, only useful to bail out big financial institutions and banks. The original source of the 2008 financial crisis—cheap credit and lax lending standards that fueled a housing bubble and put the loans at high risk of default—showed how vulnerable our lives truly are, that reckless speculations can destroy our being in a matter of few years. As Alain Badiou states, "We must reverse the old verdict according to which we would be in '*the end of ideologies*'. We see very clearly today that this so-called end has no other reality than the slogan 'save the banks'" [26].

Here, it is important to reread the works of Antonio Gramsci, the Italian political philosopher. Gramsci's great insight was that in the face of changes that could wipe out the world as we know it, a new form of thinking and political act is needed to abolish the old regime:

> Events are the real dialectics of history. They transcend all arguments, all personal judgments, all vague and irresponsible wishes. Events, with the inexorable logic of their development, give the worker and peasant masses, who are conscious of their destiny, these lessons. The

> class struggle at a certain moment reaches a stage in which the proletariat no longer finds in bourgeois legality, i.e., in the bourgeois State apparatus (armed forces, courts, administration), the elementary guarantee and defense of its elementary right to life, to freedom, to personal safety, to daily bread. It is then forced to create its own legality, to create its own apparatus of resistance and defense. [27, p. 17]

Referring to Italy's experience during *Risorgimento*, Gramsci listed two types of revolution: the active revolution led by Giuseppe Mazzini and the passive revolution led by Camillo Benso di Cavour. The passive revolution entails an attempt to create cultural hegemony and change through a meaningful process that, from Gramsci's perspective, would be achieved by patiently preparing for radical and revolutionary transformations. Today, in the context of our inherent vulnerability as well as the factors that cause it in the framework of politics and economy, there is a need for that same kind of comprehensive, global, and concrete project (to address everything from the political and economic crises to the ecological one).

However, what modes of emancipatory act should be applied here? One of the usual tricks of the ruling system in facing the concept of vulnerability is responsibilization. Accordingly, each of us is responsible for vulnerability and must do our duty towards it, from various forms of charity to individual responsibilities to improve the environment (recycling newspapers at home and work, etc.). Therefore, if a person is vulnerable to waves of unemployment, global infectious diseases or environmental crisis, the responsibility is not the authorities and the ruling class because of a "magical voluntarism" that is "both an effect and a cause of the currently historically low level of class consciousness. It is the flipside of depression—whose underlying conviction is that we are all uniquely responsible for our own misery and therefore deserve it" [28]. Therefore, as far as the category of vulnerability is concerned, politics in most cases acts in a reactionary way, and the only ones who are not blamed are those who are directly responsible for the formation of a vulnerable situation. In regard to preventing disaster and a vulnerable situation, our options are not only political but fundamentally ontological. Alternatively, as Jean-Luc Nancy once put it, "it is not a question of *preventing calamities*: it is a question of changing our humanity" [29].

The vulnerability elicited by the pandemic serves as the best *point de capiton* for politics since it places more emphasis on the question of the state's credibility. Accordingly, the state should not only take a much more active role and organize the production of essential items such as masks, test kits and respirators, confiscate hotels and other resorts, ensure the living conditions of those who have recently become unemployed, and so on. The state should do all this by abandoning the mechanisms of the market. However, attempts to overcome the uncertain and vulnerable situation of COVID-19 should not be confused with pastoral views that are trying to return to order and the good old days, a world Stefan Zweig once described as a world in which "everything had its norms, its definite measure and weight" [30, p. 1]. It is essential to emphasize that it is not possible to return to the good old days!

As far as today's relations in the world are concerned, one should mention lives that are disposable and deportable and the abandoning of individuals to naked forces of the market. Under capitalism, the problem is not a few rotten apples but the fact that the entire logic of the system initiates vulnerability and disposability. This vulnerability consists of a continuity between those who are poor and those who are not, which is established precisely through those who stand at risk of becoming poor. Accordingly, the younger generation now realizes that they have no use, namely, there would be in advance no jobs and even a decent place in society for them! Vulnerable and disposable life signifies a new form of apartheid in which there is an invisible gap between *inside* and *outside.* The paradox we should bear in mind is that global capitalism is of course *universal,* but the price is vast invisibility of many parts of it and ultimate exclusion, with new walls emerging between the included and excluded. Consequently, as Sloterdijk elaborates, "the world interior of capital is not an agora or a trade fair beneath the open sky, but rather a hothouse that has drawn inwards everything that was once on the outside" [31, p. 12]. This interior, built on capitalist excesses, determines everything: After the process that transformed the world into the globe, "social life could only take place in an expanded interior, a domestically and artificially climatized inner space" [31, p. 171]. Here, one can even take advantage of the notion of "horrorism" that Adriana Cavarero uses to characterize the disfiguration of the body in suicide attacks: "What is at stake is not the end of a human life but the human condition itself, as incarnated in the singularity of vulnerable bodies" [32, p. 8]. The real danger of global capitalism is not only the extent of political and social inequalities, but the main point is that in such a framework, each body is intrinsically and a priori vulnerable and subject to destruction. In other words, shared vulnerability and interdependency becomes a constitutive function for forms of social, political, and economic hierarchy.

Hence, the vulnerables are paradoxically visible in the sense that their very exclusion is their mode of inclusion. Through his/her vulnerability, the vulnerable can understand the nature of the existing antagonism, and this can become a factor in taking steps to remove these obstacles. This situation should not be considered a kind of Buddhist enlightenment about the necessity of suffering to reach salvation; the nature of a crisis does not necessarily lead to a deeper understanding, but the inconsistency of a situation, its essential dislocation/exclusion is constitutive of the situation itself, i.e., the position of universality embodied in the excluded.

In general, until now, our collective responses to an Armageddon have been similar to grief's stages: ideological denial, explosions of anger and attempts at bargaining, followed by depression and withdrawal. However, is that our only option? Vulnerable to ourselves, after passing through this zero-point, we can begin to perceive the crisis as a chance for a new beginning. This New Beginning, far from being an organic community or an authoritarian communitarian regime, should be built upon a new substantial community that is an egalitarian collective that puts an end to the current apocalyptic proletarianization.

5 The Question of Commons

If vulnerability is our fate, how should the vulnerable subject be defined, a subject caught between liberal democracies as the best of the worst and other options? The importance of asking the right question is evident here. We can take advantage of our fundamental vulnerability to expose false options and emphasize our critical nature, which made us vulnerable in the first place. One of the most critical issues for the Left in this situation is to focus on the real enemy: the reason for our vulnerability and degradation as a species. Rather than finding solutions, pandering to convenient options only creates confusion, which hides the very essence of decline. In other words, it should be emphasized without any shame or doubt that "the principal problem of capitalism is not in neoliberalism, or in austerity politics, nor in new forms neither of authoritarian or apartheid regimes, nor in sexism, homophobia or racism, and so on, but in the capitalist form itself, that is, in the *value form*. Instead of referring to neoliberalism as the cause for our plights and miserability, we should (at the risk of sounding archaic), bring back the critique and the overcoming of *capital* as the ultimate goal of our thinking and actions" [33]. In other words, one can say that human vulnerability to various types of current disasters (recent health emergency, threat of neoliberal exploitation, new forms of apartheid) as well as disasters that we will witness in the near future (in other words, the full-scale environmental disaster of which we are still witnessing prototypes) is nothing but the biggest market failure in human history.

There are many examples of this failure. A report by the World Bank has stated that heat waves in India could break the human survivability limit in the coming years. In a report published by the World Bank, India could experience heat waves that could make human survival impossible. The country has recorded several deaths due to intense heat waves over the last couple of decades. The World Bank report stated that the country is experiencing higher temperatures that arrive earlier and stay far longer, Business Insider reported [34]. Accordingly, in April 2022, India was plunged into the grip of a punishing early spring heat wave that brought the country to a standstill, with temperatures in the capital, New Delhi, topping 46 °C (114 °F). The month of March, which witnessed extraordinary spikes in temperatures, was the hottest ever recorded. It is interesting to note that the warning is of a different nature in this context; the World Bank does not refer to a global catastrophe threatening millions of people and is undoubtedly just the beginning of Armageddon's environmental dominoes but rather as "Climate Investment Opportunities in Indian Cooling Sector". If this paradox does not reflect our increasing vulnerability to future disasters, it cannot have any meaning in today's world.

Thus, the one and only question today before us is as follows: do we endorse the predominant naturalization of capitalism, or does today's global capitalism contain antagonisms that are sufficiently strong to prevent its indefinite reproduction? What Michael Hardt and Antonio Negri call the "commons", the shared substance of our social being, is of absolute necessity here [35, p. viii]. The important point in this regard is to try to preserve these commons and turn them into a battlefield of vulnerabilities. This enables us to see the progressive "enclosure" of the commons as a process of proletarianization of those who are excluded from their own substance and declared vulnerable.

In her *The Corona Crash: How the Pandemic Will Change Capitalism*, Grace Blakeley challenges the idea that because the pandemic is being associated with an increase in state spending, it heralds the end of neoliberalism. She argues that neoliberalism was never truly about shrinking the state; it represented a reorientation of state power away from a corporatist model, in which the interests of workers were balanced with those of capital, towards a financialized model, in which the interests of capital—and financial capital in particular—are placed at the center of economic policy [36]. According to her, the result of the pandemic is therefore going to be a much more centralized economy—one in which a small number of firms, financial institutions and states survive from the carnage created by this pandemic, giving them a great deal more power over society: "Rather than a free market economy, we're living in a system of 'planned capitalism', in which a small number of very powerful institutions control who gets what" [37]. Here, the usual readings and definitions of capitalism in the twentieth century are no longer sufficient by themselves. Today, we are facing a more brutal form of the authoritarian regime, namely, ultracapitalism. In ultracapitalism, we are not faced with classic exploitation but rather the privatization of what Marx called the common good. A couple of mega-companies will control everything and be in cahoots with the state security apparatus as capitalism becomes more feudal and digital. In this situation in which our commons are in danger, everyone is by default vulnerable.

Žižek insists that war and ecological threats are even more menacing than the pandemic, and for much of humanity are experienced as such. This emphasis means that the pandemic is not a random event in the global arena but the result of a system based on internal/external logic, a system that creates a new disaster for mankind every few decades, a system whose usual operation is based on the vulnerability of a large number of those who live inside or outside its borders. In his view, we keep imagining a return to normal that is not possible and therefore are in dire need of a new conception of normality. As Žižek suggests, we have to accept that returning to a nonvulnerable state of being, a return to normal order of things, is impossible, at least as long as global capitalism continues to reign. Our real goal should be to introduce a new concept of normality and vulnerability and reflect on human nature and our basic stance toward human life. As Žižek writes in his *Pandemic!2 Chronicles of Time Lost*, "the secret wish of us all, what we think about all the time, is only one thing: when will it end? However, it will not end: it is reasonable to see the ongoing pandemic as announcing a new era of ecological troubles" [38, p. 12]. This means that the state of vulnerability is not a temporary status but an ongoing global "standard of living", only a precursor to future pandemics and instabilities. This new situation inevitably creates new forms of the concept of class and proletariat that may not have a place in the classical framework of Marxism. Much of this class is not exploited in the classic Marxist sense of working for those who own the means of production; they are exploited in the way they relate to the material conditions of their life, i.e., through their basic vulnerability. To put it more specifically, "risk and vulnerability generate new forms through which both populations and individuals are governed through the mere potential (or alleged potential) for harm" [39, p. 72]. This function of vulnerability representation serves to critique existing policy, and it can be a powerful tool to alter existing forms of government.

One can even take a more radical stance in here. To use the terminology of Jean-Luc Nancy, today, we need to hold a "deconstructionist" approach to the world. As Shaj Mohan recently elaborated, by definition, deconstruction created two processes that gave reason to each other: both the revelation of the polynomia—by which we understand the ability of "things" to create many regular forms—that constituted the identity of these systems of norms and the breaking of the levee of the systems that held off the new homological powers to release these powers into the currents of the world [40]. This form of deconstructionist is an absolute necessity in our vulnerable situation, which enables us "to take apart, to disassemble, to loosen the assembled structure in order to give some play to the possibility from which it emerged" [41, p. 148]. This form of deconstruction can prompt new forms of inclusive ethics and empathy, as well as a renewed emphasis on the importance of such approaches. We should capitalize on, rather than manipulate, this surge of solidarity and must have faith in the possibilities of this new solidarity.

In *Totem and Taboo*, Freud explains that the attainment of kingship and community leadership in ancient times was not something of pleasure and a source of happiness but rather a source of suffering; the ruler bears such a heavy responsibility that in most cases he clearly favored someone else to take his place:

> The dignity of their position ceased to be an enviable thing, and those who were offered it often took every possible means of escaping it. Thus in Cambodia, where there are kingships of Fire and Water, it is often necessary to force successors into accepting these distinctions... The savage Timmes of Sierra Leone, we learn from Frazer, who elect their king, reserve to themselves the right of beating him on the eve of his coronation; and they avail themselves of this constitutional privilege with such hearty goodwill that sometimes the unhappy monarch does not long survive his elevation to the throne. [42, pp. 54–58]

This is exactly what we need right now. We must once more rekindle this old logic in light of the numerous disasters that surround us, from the threat of an impending environmental catastrophe to pandemics and geopolitical conflicts, to make our rulers afraid to put the lives of thousands of people in danger and normalize vulnerability. The ancient method of the natives of Sierra Leone shows that disasters such as the pandemic are the best opportunity to question the mechanisms of governments and try to fundamentally change the existing geopolitical and political situation. In other words, this is the political achievement of the pandemic, if any; our vulnerability is the core of our demand for the impossible, namely, abandoning the mechanism of the market and changing the very circumstances that make us vulnerable.

Core Messages

- Our vulnerability is the core of our demand for the impossible, namely, abandoning the mechanism of the market and changing the very circumstances that make us vulnerable.

- We have to try to preserve our commons and turn them into a battlefield of vulnerabilities. This enables us to see the progressive "enclosure" of the commons as a process of proletarianization of those who are excluded from their own substance and declared vulnerable.
- Acceptance of fundamental vulnerability as a substantive negativity is the key to overcoming the predicament we are all in.

References

1. Murakami H (2011) 1Q84, trans. Rubin J, Gabriel P. Knopf, New York
2. Baudrillard J (1994) The illusion of the end, trans. Turner C. Polity Press, Cambridge
3. Giuffrida A, Tondo L (2020) Singing stops in Italy as fear and social unrest mount. Guardian, April 1
4. Marx K (1992) A contribution to the critique of Hegel's philosophy of right. In: Early writings, trans. Benton G, Livingstone R. Penguin Publishing Group, London
5. Laporte D (2002) History of shit, trans. Benabid N, El-Khoury R. MIT Press, Cambridge
6. Mohan S, Dwivedi D (2019) Gandhi and philosophy: on theological anti-politics. Bloomsbury Publishing, London
7. Freud S (2000) The interpretation of dreams, trans. Brill AA. Wordsworth, Stansted
8. Žižek S (2008) The sublime object of ideology. Verso, New York/London
9. Timofeeva O (2014) The end of the world: from apocalypse to the end of history and back. e-flux J, June
10. Dupuy J-P (2015) A short treatise on the metaphysics of tsunamis. Michigan State University, East Lansing
11. Žižek S (2012) Less than nothing: Hegel and the shadow of dialectical materialism. Verso, New York/London
12. Hegel GWF (1977) Phenomenology of spirit, trans. Miller A V. Oxford University Press, Oxford
13. Mackenzie C, Rogers W, Dodds S (2014) Introduction: what is vulnerability and why does it matter for moral theory? In: Mackenzie C, Rogers W, Dodds S (eds) Vulnerability: new ethics and feminist philosophy. Oxford University Press, Oxford, pp 7–9
14. Butler J (2016) Rethinking vulnerability and resistance. In: Sabsay L, Butler J Gambetti Z (eds) Vulnerability in resistance. Duke University Press, Durham, pp 12–27
15. Butler J (2004) Undoing gender. Routledge, London
16. Achella S (2021) The dark side of thought. the body, the unconscious and madness in Hegel's philosophy. In: Achella S, Iannelli F, Baptist G, Feloj S, Li Vigni F, Melica C (eds) The owl's flight; Hegel's legacy to contemporary philosophy. De Gruyter, Berlin, pp 23–37
17. Žižek S (2001) Enjoy Your Symptom! Jacques Lacan in Hollywood and Out. Routledge, London
18. Derrida J (2005) The politics of friendship, trans. Collins G. Verso, New York/London
19. Graeber D (2021) After the pandemic, we can't go back to sleep. Jacobin, October
20. Watts J (2020) Arctic methane deposits 'starting to release', scientists say. Guardian, October 27
21. Anthony A (2019) Srećko Horvat: 'The current system is more violent than any revolution'. Guardian, April 21
22. Baradaran K (2021) The ruins of our lives: a plea for fatalist sleeplessness. The Philosophical Salon, August 9
23. Castel R (2003) L'insécurité sociale. Qu'est ce qu'être protégé? Seuil, Paris
24. Ghandeharian S, FitzGerald M (2021) COVID-19, the trauma of the 'real' and the political import of vulnerability. Int J Care Caring 6(1–2):33–47

25. Rancière J (2010) Ten theses on politics. In: Corcoran S (ed) Dissensus; on politics and aesthetics. Continuum International Publishing Group, London/New York, pp 27–44
26. Badiou A (2008) De quel réel cette crise est-elle le spectacle? Le monde, October 17
27. Gramsci A (1977) Real dialectics. In: Gramsci A (ed) Selections from political writings 1921–1926, ed. and trans. Hoare Q. Lawrence and Wishart, London, pp 15–16
28. Fisher M (2014) Good for nothing. the occupied Times of London, March 19, available at https://theoccupiedtimes.org/?p=12841
29. Nancy J-L (2021) To be listening; interview with Kamran Baradaran. Philosophy-world-democracy, March
30. Zweig S (1964) The world of yesterday: an autobiography. University of Nebraska Press, Lincoln
31. Sloterdijk P (2013) In the world interior of capital: towards a philosophical theory of globalization. Polity Press, Cambridge
32. Cavarero A (2009) Horrorism: naming contemporary violence, trans. McCuaig W. Columbia University Press, New York
33. Hamza A (2019) The problem of the left is its reactive position in politics; interview with Kamran Baradaran. Iranian Labor News Agency (ILNA), May
34. Jankowicz M (2022) India could suffer heatwaves that 'break the human survivability limit' World Bank says. Business Insider, December 9
35. Michael H, Negri A (2009) Commonwealth. Harvard University Press, Cambridge
36. Blakeley G (2020) The corona crash: how the pandemic will change capitalism. Verso, New York/London
37. Blakeley G (2021) Sleepwalking into a world without a future; interview with Kamran Baradaran. Iran's Book News Agency (IBNA), April
38. Žižek S (2020) Pandemic! 2 chronicles of time lost. OR Books, New York
39. Kaye K, Amuchástegui A, George A, Navarro T (2012) Neoliberal vulnerability and the vulnerability of neoliberalism. In: Bernstein E, Jakobsen JR (eds) Paradoxes of neoliberalism: sex, gender and possibilities for justice. Taylor & Francis, London, pp 71–108
40. Mohan S (2022) Deconstruction and anastasis. Qui Parle 31:339–344, December 1
41. Nancy J-L (2008) Dis-enclosure: the deconstruction of christianity, trans. Bergo, B, Malenfant G, Smith MB. Fordham University Press, New York
42. Freud S (2004) Totem and taboo, trans. Strachey. Routledge, London

Kamran Baradaran is an Iranian translator, author, critic, musician, journalist and associated editor at Philosophy-World-Democracy journal. Among others he has translated "Demanding the Impossible", "Against the Double Blackmail: Refugees, Terror and Other Troubles with the Neighbors", "First As Tragedy, Then As Farce", and "Pandemic!: COVID-19 Shakes the World" by Slavoj Žižek; "Art and Fear" by Paul Virilio; "The Spirit of Terrorism" and "Fragments: Interviews with Jean Baudrillard" by Jean Baudrillard; two collections of Antonio Gramsci's pre-prison writings, into Persian. He has also written the book: *Feminine Writing: Improvisation in the Mist* (Rowzaneh Publishing House, Tehran, 2016). He works as a senior journalist specializing in philosophy, critical thinking, literature and international politics and his articles have appeared in Iran and around the world. He interviewed Slavoj Žižek, Divya Dwivedi, Simon Hajdini, Agon Hamza, Adrian Johnston, Shaj Mohan, Robert Pfaller, Alan Shandro, and Guy Standing which have been published in different journals and newspapers. He is currently working on a book in collaboration with Anthony Ballas, focusing on the rise of the far-right and the redefining of the Left.

Chapter 13
Pandemic Necropolitics: Vulnerability, Resilience, and the Crisis of Marginalization in the Liberal Democratic State

Ubaka Ogbogu

> *... sovereignty consists in the power to manufacture an entire crowd of people who specifically live at the edge of life, or even on its outer edge—people for whom living means continually standing up to death, and doing so under conditions in which death itself increasingly tends to become spectral... As a rule, such death is something to which nobody feels any obligation to respond... Necropolitical power proceeds by a sort of inversion between life and death, as if life was merely death's medium.*
> *Achille Mbembe, Necropolitics, 2019*

Abstract Vulnerability, marginalization, and resilience in the pandemic and in an eventual "post-pandemic" state are examined through the lens of Achille Mbembe's theory of necropolitics. The central claim made is that vulnerability and marginalization are products of a covert and intentional politics of death. It is also argued that for the vulnerable and marginalized, the pandemic does not demarcate between a previous normal and eventual normal state, but is rather, an escalation of a persistently abnormal state. A final claim is that reflection on the fate of the vulnerable and marginalized must resist a Kantian impulse to find and urge resilience and focus instead on a direct attack on the necropolitics that sustains suffering for this population.

Keywords COVID-19 pandemic · Necropolitics · Vulnerability · Resilience · Marginalization · Solidarity

---

U. Ogbogu (✉)
Faculty of Law, University of Alberta, Edmonton, Canada
e-mail: uogbogu@ualberta.ca

S. Achella and C. Marazia (eds.), *Vulnerabilities*, Integrated Science 18,
https://doi.org/10.1007/978-3-031-39378-5_13

1 Introduction

The COVID-19 pandemic laid bare the tragedy of vulnerability *and* marginalization (viewed as a singular conjunctive construct) that has long been the calling card of the liberal democratic state. That the pandemic occurred, and that it had differentially adverse impacts on the already doomed, are not merely accidental, but rather inevitabilities inherent in and fostered by the verticality of power relations within the liberal state. Here, vulnerability is immanent in the positionality and relations of the marginalized to the powers constituted in the state against them, such that one is but not merely a synonym of the other, but are, in fact, one and the same. The survival of the vulnerable-marginalized in this state of affairs is therefore, first and foremost, an act of resilience, while also futile—both in scale and effect—against the necropolitical constitution and aspirations of the state.

By examining the "nonlaw laws", i.e., "the law that originates in nonlaw and that is instituted as law outside the law," [1, p. 27] and various acts of executive violence and indifference through which the pandemic was weaponized against the vulnerable-marginalized, this essay situates pandemic public health governance in the liberal democratic state as both an extension and realization of necropolitics. Achille Mbembe's powerful and evocative political theory, founded in the social and political organization as "power and the capacity to dictate who may live and who must die," [1, p. 66] is obvious as an explanatory lens, as what follows shows. More importantly, it serves to redirect attention and emphasis from misfortune and irresponsibility to design and inevitability, as primary frames through which one must see how so many came (and continue) to die in the pandemic. For in these stories of necropolitics, one observes default premeditation regarding who must die, such that the state's response to the suffering wrought by the pandemic must be, by design, merely a realization of that which is inevitable. They, *who must and did die*—the poor, the racialized poor, seniors in care, the disabled, the migrant worker, the *essential* worker, the globalized poor—are already constituted and doomed, such that what pandemic policy demanded was a realization of their doom rather than their salvation or resilience.

But within this state of affairs, the situation that these doomed (the vulnerable-marginalized) find themselves—that is, in *continually standing up to death*---is one that is endogenously resilient, albeit interspersed with a hopelessness that they are powerless to undermine or change. In this matrix of hopelessness and resilience lies opportunities for a redemptive default in an eventual post-pandemic state (if such a state was indeed possible), one in which their vulnerability and the endogenous resilience becomes itself an act of resistance against an ordained and decidedly hopeless set of outcomes.

The fundamental point, therefore, and the thesis asserted is, that for the vulnerable-marginalized, there is no post pandemic state. Rather, there are states of escalation of and return to a default *abnormal state* that defines and embodies their existence and inexistence within a polity that is ordered and designed to kill them, or to procure their misery and eventual death. In this sense, to speak of a post-pandemic state is to engage a literary reality that assumes a normality absent in the pandemic state, or at least, some qualitatively different state of affairs from a pandemic state. The lived reality of the vulnerable-marginalized is not consistent with these literal distinctions; rather, it is characterized by an always "spectral" engagement with misery and death that is sometimes deepened by more exogenous misery or death, or misery and death causing factors.

2 The Abnormal State

In conventional public health language, we have come to understand this abnormal state that the vulnerable-marginalized exist in as chiefly defined by the absence of the determinants of health. This characterization, while empirically supportable, presumes that the abnormal effect is remediable by providing for such determinants. However, even so, causality in the abnormal state is incomplete without engaging with its two principal agents, namely the positivistic (in the sense of practice over ideal) [2] legal and political program of disenfranchisement (in the totality of lived and abstracted experience), and the normalization of inequity. Both operate as a unified construct; the former serves as the political expression of the latter, propped up by an *indecently indifferent*[1] propertied class whose acceptance of life as inherently uneven is both dull and deadly. Their unity is also consonant with the necropolitical inversion of democracy to serve the "generalized cheapening of the price of life and...habituation to loss" [1, p. 38].

Within the abnormal state, the *haves* (whether by virtue of birth, race, striving, accident, capitalistic greed, intergenerational inheritance, or any of the other permutations that produce wealth, and sometimes concomitantly, marginalization and suffering) strive to become *have-mores*, thus creating a polity that is defined and designed by reference to the wants and perspectives of the invulnerable and privileged. Within this state, not *having* is boring and listless, contorted (or relegated) as it were to reside in projects of hope and public interest advocacy by tireless but often-ineffective social justice warriors. This reflection is evident in decades of social policy in the Western hemisphere focused on creating and sustaining "a strong middle class," a strange concept if ever there was one, that presumes the existence of an inevitable (or, at least, unassailable) "high class," and an inevitable but adjectivally and realistically unfortunate low class. This longstanding commitment to creating and fostering a middle class, which holds across political and ideological divides, also operates to immobilize a "looking back" at who has been left behind.[2]

Vulnerability, especially the perceived kind, is of course agnostic to class. Infectious disease outbreaks often do not discriminate between classes, thus creating a sense of vulnerability that transcends the presence or absence of material needs. Still, a deeper sense and reality of vulnerability subsists in the *absence of material needs*,

[1] I am grateful to Shree Paradkar for the italicized phrase.

[2] I owe this insight to my colleague Godwin Dzah.

which, as the COVID-19 pandemic has shown, intersects with marginalization, race, economic status, and geography [3, 4]. In the abnormal state, the perceived vulnerability of the privileged classes drives policy, including public health policy. Take, for example, government supports during the pandemic, which were largely premised on employment, home ownership, and business ownership, also the usual demarcations between privilege and marginalization in liberal democracies. Or, worse, the crass resumption of "normal life" through the abandonment of the public health protections to the detriment of the immune-compromised. The political and practical humiliation of the vulnerable-marginalized within the abnormal state, therefore, subsists in deterministic forms of material and political neglect that the privileged classes can neither perceive nor experience.

There is another sense in which the abnormal state intersects with necropolitical power, to wit, how both posit *natural* catastrophe as an inevitability rather than as creations of statecraft. In this way, catastrophe is causally externalized, and the state's engagement with it is then defined and examined through the lens of the unwitting shield and guardian of ordered existence versus disordered fortuity. Extempore, and putting aside the contested theories about gain-of-function viral escape from a Chinese laboratory, the COVID-19 pandemic appears to be, and has been presented as, an accidental existential disruption.

Nothing could be further from the truth. Underinvestment in surveillance programs for emerging zoonotic disease threats is a specific governmental policy that invites a degree of culpability in the creation of a pandemic than those responsible for it would care to admit [5]. Perhaps more sinister is the fact that, for the necropolitical state, the pandemic immediately presented an opportunity to entrench its grip on the power to dispense or facilitate death under the façade of "crisis management." This claim is evident in the fact that with millions of deaths and counting, the pandemic has not forced a reckoning with the silent political and social forces that make death inevitable for the marginalized in a necropolitical state, including racism, poverty (as a state of being), disability, carcerality, and systemic misogyny. Rather, the earliest (and to date, most) pandemic management policies have focused on threats posed to the economically-empowered body politic. These include policies that protect the privileged nationalistic and/or itinerant body (travel bans and restrictions, quarantine, vaccines, proof of vaccination), the economically-capable body (masking, ventilation, work from home), and, conversely, those that entrench the misery of the marginalized body, as is the case with the paradoxical "personal responsibility" policies, which purport to treat social responsibility and public health as matters to be decided and handled by individuals. Thus, both before and during this pandemic, the necropolitical state is constituted to "manufacture an entire crowd of people who live specifically at the edge of life… [and for] whom living means continually standing up to death" [6, p. 37].

Consider two examples.[3] On January 8, 2021, the government of the province of Quebec in Canada, in a bid to control the rising number of COVID-19 cases in the province, issued a curfew decree that prohibited "any person" from being outside "their residence or what takes its place or the land of such residence" between 8 p.m. and 5 a.m. Contravention of the decree attracted a fine ranging from $1000 to $6000. In issuing the decree, the government inexplicably failed to consider the obvious impact on persons experiencing houselessness, especially given that the decree came into force in one of the coldest months of the Winter season. While at first glance this seemed like an inadvertent omission, subsequent events suggested a callous disregard for the affected houseless population amounting, at a minimum, to culpable commission. Specifically, various individuals and groups, including the Mayor of Montreal (Quebec's largest city), organizations providing houselessness relief, and public interest lawyers representing those experiencing houselessness, alerted the government to the discriminatory and disproportionate impact that the decree would have on people experiencing houselessness. The government, unmoved, refused to allow an exception for this population, thus forcing a group of legal aid lawyers working on behalf of houseless and indigent clients to petition for a court-ordered exception [1, p. 37].

In *Clinique juridique itinérante c. Procureur général du Québec*, the court ruled in favour of the houseless and suspended the decree as it applied to homeless people [7]. The court agreed with the petitioners that the decree "would undermine the rights to life, liberty and security" of homeless people as protected under the Canadian and Quebec charters of rights and freedoms, would have a discriminatory and disproportionate effect on them, and would cause them irreparable harm if implemented. Strangely, the government contended that the measure was not intended to apply to homeless people "for whom it would be impossible to have access to a shelter." The court rejected the government's position because it presented an enforcement impossibility, viz., that a homeless person could not know in advance if the police, in enforcing the curfew, would consider that it was impossible to have access to a shelter or exercise their discretion to tolerate this exception. However, as the petitioners pointed out, the aim of forcing the homeless into shelters was itself fraught with problems, considering that they were at a higher risk of contracting COVID-19 in the typically crowded shelters, and many among them, even having access to a shelter, tend to leave because they have addiction and mental health problems that

[3] To avoid decontextualizing examples from other jurisdictions, I discuss only Canadian examples. For a recent discussion of some U.K. examples, see [3]. See also *Regina (Gardner and another) v Secretary of State for Health and Social Care and others* [2022] EWHC 967 (Admin): Government Covid-19 directives aimed at alleviating the strain on hospitals resulted in the negligent relocation of asymptomatic patients into care homes, without quarantine. Residents in care homes (generally the elderly) were infected and died from the virus; BVerfG, Order of the First Senate of 16 December 2021—1 BvR 1541/20—, paras. 1–131, http://www.bverfg.de/e/rs20211216_1bvr154120en.html, where Germany's Federal Constitutional Court ruled that the Legislature (*Bundestag*) had violated the Basic Law (*Grundgesetz*) by failing to enact policies that would protect persons with disabilities from treatment discrimination in the likely event of shortages of life-saving intensive care resources during the pandemic.

are either not tolerated in shelters or that make it untenable to exist comfortably in such spaces.

This case provides an apposite lens into the mind and workings of the necropolitical state. It is difficult to view the actions of the Quebec government as suggesting anything other than an acceptance that houseless persons are entirely disposable, even as they ostensibly pursue the well-intentioned goal of preventing the further spread of a deadly pandemic. More damningly, the government purposefully missed an opportunity to engage with houselessness as a social issue that should not exist in the first place, instead choosing to view houselessness as a *normalized* and *abstracted* imaginary rather than the actual reality of the abnormal state in which those facing it are forced to subsist in.

For the houseless population implicated by the government decree, it is striking that the limit of the government's redemptive imagination, in the context of a pandemic, is a shelter (one suspects that this would also be the case in a so-called post-pandemic state). The shelter is a symbol of the sum of the liberal democratic state's exertions towards the least fortunate in society—a place that may provide a brief respite from biting circumstances, but which renders those for whom it is made available even more subject to the vulnerabilities that are ingrained in their existence within the state.

A second example derives from a pair of directives issued by agents of the government of the Canadian province of Alberta during a recent wave of the COVID-19 pandemic. The first, issued by the province's Chief Medical Officer of Health (CMOH), rescinded a previous order requiring masking in elementary and high schools [8]. The second, from the Minister of Education, purported to forbid school boards from imposing their own masking requirements [9]. In the context of a pandemic where there have been incessant and often vexing debates regarding the extent of its impact on children, both actions rang as particularly callous towards children who were immune-compromised and at heightened risk of severe outcomes if they contracted COVID-19. The directives forced these children and their guardians into a cruel *non-choice*: skip school or risk contracting a disease that could prove deadly.

A court challenge brought by and on behalf of the affected children provides insight into the necropolitical motivations of the state actors involved [10]. The challenge focused on two matters; the reasonableness of the order issued by the CMOH, and the legal effect of the ministerial directive, which issued by way of a letter addressed to school boards.

Regarding the CMOH order, the court found that it was "an unreasonable interpretation of the *Public Health Act*" because the CMOH did not in fact make the order as commanded by the Act, but instead "merely implemented a decision of a committee of Cabinet." The court was unable to determine the basis for the decision by the Cabinet committee from the evidence proffered—indeed, the reasons remain opaque, although evidence considered but not admitted in the case suggests that the decision was based on "political considerations" and the bizarre, unproven claim that masks have harmful effects on children's cognitive and emotional development [11]. Even if one accepts these considerations, it is difficult to imagine how they possibly supersede the welfare of the complainants in the case, who stood to suffer the most from the government policy, unless of course if one's imagination accepts their suffering as inevitable or necessary for the sake of political expediency.

The Minister's letter is even more troubling. It repeated the questionable claim that there are "mental health impacts that come along with public health measures such as masking" and directed that school authorities "cannot deny their students access to in person education due to their personal decision to wear or not wear a mask in schools." Much like the Cabinet committee's decision, the letter failed to mention or consider the welfare of immune-compromised school children, choosing instead to urge and rely on "many important factors to consider" such as "seeing the facial expressions of teachers and classmates" and "having the ability to be animated and joyful" as reasons for discarding the masking requirement. It is again hard to imagine how these factors can trump the risk of severe harm posed to the applicants, unless one considers that enhancing the joyfulness of the general school population is more necessary than protecting those for whom life has become merely death's medium.

The court found that the Minister's letter had no legal effect because it was issued by way of a ministerial letter rather than by a formal regulation. This conclusion, which is also reflected in the opaqueness of the Cabinet committee's decision that informed the CMOH order, highlights the *modus operandi* of the necropolitical mind, which is a tendency towards *covert, inscrutable or legally immune* reasons and actions. This approach renders the law ineffective in vindicating the rights and interests of those affected, either because the issues raised are non-justiciable, or because legal intervention (assuming that the affected have the means to pursue it) arrives too late. Indeed, the court decision did not reinstate the masking requirement since a subsequent order by the CMOH rescinded all extant public health orders.

What these two examples share in common is a tendency towards crisis management that serves those privileged enough to comply with the management directives, or, at least, privileged enough to regard the directives as mere disruptions of an ordered or (in the context of a pandemic), a semi-ordered reality. A curfew might be a legitimate public health order for those who have the means and residence to retreat from the streets, but it is a deepening or escalation of marginalization and despair for the homeless. A decision to rescind a mask requirement might serve the interests of those prepared to move on from a still raging pandemic, but it is effectively a death sentence for the immune-compromised.

Observing this duality of crisis management for some and crisis deepening for others recalls Saptarishi Bandopadhyay's powerful maxim, that "modern governance nurtures order and disorder in pursuit of idealized visions of prosperity and rule at the limits of nature and society" [12, p. XIII]. What Bandopadhyay so painstakingly makes clear in his ground-breaking work is that in liberal democratic states, disorder and catastrophe are rarely innocent, but instead are products of a "muscular-neoliberalism" that serves and is often applauded by an indifferent and privileged middle- and upper-class citizenry, because the illusion protects their comfortable belief that "states exist to protect society from the inevitable occurrence of disasters"

[12, p. XIV]. In essence, the duality underscores a suspicion that the state repeatedly rationalizes the occurrence of catastrophic events, such as the pandemic, by securing agreement from a powerful segment of society that the destruction wrought by disease is *the normal order of things*. Thus, death, which is *terminal* to those who did not participate in but are affected by this agreement, becomes a *normal pandemic-related condition* which is within the *normal order of things*, and not an abnormal occurrence.[4]

3 Non-law and Covert Necropolitics

Contrary to the examples discussed in the preceding section, many (if not most) of the liberal democratic state's necropolitical dealings cannot be vindicated through legal or judicial means. These necropolitical projects, defy common knowledge because they are published as casual statements rather than rules of law, and often involve *enforceable* actions, directives and reasons that are purposefully non-justiciable. In this way, they are soffits to a state apparatus of non-legality and covert pronouncements of immoral law [13]. Still, the fact that legal or judicial vindication is present is some cases is not to be taken as indicative of a trend, or as suggesting that such vindication, for the marginalized, is necessarily available or *vindicative*. Rather, legal and judicial processes, which the marginalized can only access through legal aid and public interest litigation, are, at best, projects of hope that depend deeply on the benevolence of private and state actors. Interestingly, the question of how the marginalized can maintain predicable access to legal vindication of their rights and interests, or regarding whether constitutionally-grounded rights guarantees are meaningful without such access, has been and remains a niche interest among legal scholars. The reasons for this neglect are not central to this paper, but it warrants mention that possible reasons lie in the legal academy's blind adherence to the theoretical possibilities of rule by or of law and the penchant for beginning and ending their examination of rights from the perspective of legal doctrine.

The pandemic is littered with evidence of these covert non-laws that operate as law. In a speech in the legislature that preceded the government's policy to remove pandemic restrictions and reopen the economy, the then Premier of Alberta, Jason Kenney, called the COVID-19 virus "an influenza" and said the following:

> We cannot continue indefinitely to impair the social and economic — as well as the mental health and physiological health of the broader population — for potentially a year for an influenza *that does not generally threaten life apart from the elderly and the immunocompromised*... The average age of death from COVID in Alberta is 83. And I remind the house the average life expectancy is 82. [14]

This stunning invocation of necropolitics captures the afore-discussed elements: the intentional manufacture of a crisis (*i.e.* a continual state of disease) in the "pursuit of idealized versions of prosperity" and for the betterment of those who already

[4] I am grateful to my colleague Godwin Dzah for this insight.

have the mental and physiological fortitude to face the crisis, and an intentional *worsening* for the living dead, whose life can be conveniently threatened for the sake of the social and economic, and because they are already considered to have lived beyond what the state can tolerate. These statements, far from the mere rantings of a possibly psychopathic leader, would emerge as law-based policies including the rescindment of masking requirements in schools, premature endemicity and a consequent relaxation of public health restrictions, and neglect of reforms proposed for the protection of seniors, especially those in congregate care, who account for the highest proportion of disease morbidity. In the same vein, post-COVID conditions (colloquially, long COVID) is a health care crisis manufactured by intentional state policies of neglect, and which created a mass disabling event that, for those already lacking access to health care and determinants of health, amounts to creating scores of the living dead.

Kenney is not alone in deploying the life of the vulnerable-marginalized as simply death's medium when proposing or justifying state action or inaction on the pandemic. As the death toll in the United States reached two hundred thousand recorded deaths, Donald Trump famously dismissed the pandemic as not worthy of a bother because it only "affects elderly people [and] elderly people with heart problems" and called for states to reopen schools because the virus "affects virtually *nobody*" [15]. Texas Lieutenant Governor Dan Patrick called upon senior citizens to "take a chance on [their] survival, in exchange for keeping the America that all America loves for [their] children and grandchildren" [16]. In Ontario, Premier Doug Ford cancelled a minimum wage increase instituted by a previous government and reduced social assistance as the pandemic picked up pace [17]. For these necropoliticians, to be *somebody* deserving of the state's attention or protection is to be among a class of persons that do not depend on the state for attention or protection. Conversely, a *nobody* is one whose existence is a mere means to non-existence, and whose inevitable death the state can foster through policy-forming rhetoric or actions that are immune from legal attack.

An even more sinister perversion, namely racism, is revealed when one peels back the layers of these deadly pronouncements. As Mbembe observes, "[t]o a large extent, racism is the driver of the necropolitical principle insofar as it stands for organized destruction, for a sacrificial economy, the functioning of which requires...a generalized cheapening of the price of life and...a habituation to loss" [1, p. 38]. Throughout the pandemic, racism towards marginalized races has been the dividing line between good health and a diseased state, life and death, and suffering and coping. Such racism is inherent in conditions created by the state whereby "certain lives have been valued, nourished, and protected while others have been devastated and destroyed" [18, p. 60]. These devastated others, the elderly, sick, immune-compromised, houseless, poor and other living dead folk, who must then be sacrificed to serve the greater good, consist disproportionately of the racialized (in Canada, Black and Indigenous peoples).

Racism also inheres in the colour-evasiveness and the pretense of non-racial causality that characterized the liberal democratic state's dithering response to the pandemic's worst impacts. Here, one sees different "racisms" at play. There is the well-documented racism of failing to collect race-based data on, and conduct race-based analysis of, pandemic impacts. The racism of excluding race as an explanatory lens, even as the suffering lines continue to be drawn by race. The racism of a disinterested occupation of the policy space by those who desire the least to help the doomed. The racism of euphemizing race through adjectival proxies that transform what Black and Indigenous peoples are experiencing into diffuse generalities. On the latter, witness, for example, how Horacio Aruda, Quebec's Public Health Director, frames the factual revelation that the pandemic has disproportionately impacted racialized persons: "it's not race that is the problem, it's the conditions of the person: poverty, crowding in houses…revenue, how many kids, university level. For me, those are the factors that can explain why those communities are more [affected]" [19] Thus, the condition of racialization is to blame, but not race that underpins the racialization itself.

What Aruda and other necropoliticians fear, is that naming race invites scrutiny and raises questions as to whether governmental neglect offends constitutional guarantees of a discrimination free life. Through words and acts of covert necropolitics, *legally enforceable commitments* against prejudice and discrimination are reconfigured as *legally benevolent necessaries*—housing, means, education, etc., that exist only at the state's pleasure (or displeasure). The *nobodies*, as perennial non-beneficiaries of the state's benevolence, are doomed to accept a cheapening of the price of their lives, and to become habituated to the loss of it, in order for the state to prosper.

In the foregoing regard, the state operates, therefore, through what Mbembe refers to as its "nocturnal face," [1, p. 27] which serves for the "exteriorization of … originary violence to third places [and] nonplaces" [1, p. 27]. By creating radical forms of power exercised without responsibility and expressed through non-justiciable reasons and rhetoric, democratic states "deaden any awareness of [the] latency" of this exteriorized violence, thus "removing any real chance of interrogating its foundations, its underneath, and the mythologies without which the order that ensures the reproduction of state democracy suddenly falters" [1, p. 27]. Thus, the political operates in anecdote, but insists that its faults be proved only through hard evidence. It defines a realm of legality that excludes what is actionable, so that it can appear in light but exist in the dark.

4 Resilience, Solidarity, and the Accidentality of Survival

Resilience is a fact of life, albeit one contingent on the presence of the *condition of living*. A condition of living is necessarily inconstant, mutable, and atavistic. Thus, one questions whether it is possible for those who are stuck in a constant, immutable and grim reality of vulnerability and death-contingent life. Much like discussions regarding the redemptive possibilities of hope, resilience ought to be treated with a healthy dose of skepticism, because it is *normally* a descriptive frame that captures the moral fixations of the privileged set rather than the actual conditions of suffering that fascinate the describer.

It is certainly the case that humans can muster resilience through sheer willpower or a survival instinct. However, this form of resilience, to the extent that it exists, is most likely a doppelganger of survival per se, or of survival against the odds. To survive a state of affairs that is ordered to kill you is not necessarily to be resilient—it could simply be happenstance or a death- and life-defying stasis akin to agony without movement.

A more useful frame for imagining and assessing resilience in the context of the grim realities of vulnerability and marginalization is socially produced forms of resilience, such as that which is built up through engagement with social institutions (family, schools, social networks) [20], or through forms or solidarity that are grounded in "practices of freedom" among self-governing, free people and "agonistic and interminable public discussions and negotiation" that transcend and defy citizenship, identity, or sovereignty [21, pp. 164–165]. But resilience in the latter senses also suffer from at least two defects, as they relate to the vulnerable-marginalized. First, participation in social institutions or in any discernible or organized in-group is doubtful for those whose disenfranchisement from society is total and enduring. The persons and communities that suffered the worst impacts during the pandemic align with the persons and communities that are excluded from the kinds of social participation that build resilience. In other words, they have never participated. Second, in a state of crisis, such as the pandemic presents, vulnerability, marginalization and social exclusion deepen and become effective barriers to participation in projects that sustain solidarity and resilience. The instinct to organize, engage, discuss and negotiate is replaced either by a survival instinct, or by a state of agonizing stasis. One can imagine that a severely handicapped person at risk of death during the pandemic as a result of necropolitical policies of neglect or purposeful harm (e.g., withdrawal of income supports) might be more concerned with survival than with the practices of freedom that foster a strengthening and revitalization of common bonds.

Solidarity could also occur through actions of "allyship" by those who possess the privilege of participation. This form of solidarity was evident, for example, in the protests that followed the forceful eviction by the city of Toronto of homeless persons and the destruction of their encampments in parks and public spaces during the pandemic [22]. The evictions were based on a judicially affirmed Toronto city bylaw that banned overnight camping in parks for the duration of the pandemic, thus deepening the vulnerability of an already vulnerable population within the context of an escalation of the abnormal state in which they routinely exist. This necropolitical action was met with protests involving shelter advocacy groups and well-meaning citizens. However, the protests were swiftly quashed through the always available and well-resourced mechanism of state violence. This suggests that allyship solidarity, though rational and moral, may be more symbolic than effective. Worse, and perhaps cynically, it could be a merely performative, thus serving to assuage the guilt of hedonistic privilege alongside the intense suffering of others. Furthermore, the extent to which such allyship actions by in-groups serve to build resilience in the out-group is unclear and possibly only contingent on the gains from the exertions of the in-group.

5 Conclusion

If my scepticism is correct, where does this take us? Hopefully, away from the illusion of, and the dogmatic idealism about, resilience, which permeates scholarly reflection on vulnerability, and towards a more direct engagement with the cancer of necropolitics. What is needed is an appreciation of the "real politics" of vulnerability and marginalization in a necropolitical state, and not the oddly sentimental Kantian impulse to seek spectral dignity from chronically hopeless conditions, all for the sake of linguistic or analytical comfort. To put it in drastically simple terms, we need the real politics of how people live, and not of how we wish them to live. As Geuss urges, political philosophy [and reflections on politics] "must start from and be concerned…not with how people ought ideally (or ought 'rationally') to act, what they ought to desire, or value, the kind of people they ought to be…but, rather, with the way the social, economic, political…institutions actually operate in some society at a given time, and what really does move human beings to act in given circumstances" [23, p. 5]. Thus, to the extent that resilience arises from how people actually live, then what ought to occupy our attention is what creates the conditions that deprive people of living, as well as how to gain resilience when there is not much living to be had.

Core Messages

- There is no "post-pandemic" for the vulnerable and marginalized. They either die or remain in a continual state of suffering and precarity.
- Slow death and precarious life during the pandemic are products of a politics of death that is purposefully undetectable and unstoppable.
- Scholarly reflection on vulnerability and marginalization in the pandemic should focus on dismantling the politics of death, rather than the resilience of those most affected.

Acknowledgements I am grateful to Godwin Dzah for insights and commentary on the paper. All errors are mine.

References

1. Mbembe A (2019) Necropolitics. Duke University Press, Durham
2. Mill JS (1865) Auguste Comte and positivism. N. Trubner & Co, London
3. Sandset T (2021) The necropolitics of COVID-19: Race, class and slow death in an ongoing pandemic. Glob Public Health 16(8–9):1411–1423
4. Dalsania AK, Fastiggi MJ, Kahlam A, Shah R, Patel K, Shiau S, Rokicki S, DallaPiazza M (2022) The relationship between social determinants of health and racial disparities in COVID-19 mortality. J Racial Ethn Health Disparities 9:288–295

5. McNeil DG Jr (2019) Scientists were hunting for the next Ebola. Now the U.S. has cut off their funding. New York Times, 25 Oct
6. Olson I (2021) Quebec curfew does not apply to homeless people. CBC News, 27 Jan
7. Clinique juridique itinérante c (2021) Procureur général du Québec, QCCS 182
8. Office of the Chief Medical Officer of Health, Record of Decision—CMOH Order 08-2022
9. Letter from Minister Adriana LaGrange to School Boards. On file with author and available upon request
10. CM v Alberta (2022) ABKB 716
11. Mckeever A (2022) Do masks really harm kids? Here's what the science says. National Geographic, 17 Feb
12. Bandopathyay S (2022) All is well: catastrophe and the making of the normal state. Oxford University Press, Oxford
13. Fuller LL (1969) The morality of law. Yale University Press, New Haven
14. Fletcher K (2020) Critics describe Premier Jason Kenney's COVID-19 speech as Trumpian. Global News, 28 May
15. Thrush G (2020) 'It affects virtually nobody,' Trump says, minimizing the effect of the coronavirus on young people as the U.S. death toll hits 200,000. New York Times, 22 Sept
16. Pittis D (2020) Politicians who consider sacrificing the old for the sake of the economy face a backlash. CBC News, 26 Mar
17. Raphael D (2019) Social murder and the Doug Ford government. Toronto Star, 7 May
18. Mawani R (2021) A historical account of the pandemic: health, colonialism and racism in Canada. In: James HF et al (eds) Impacts of COVID-19 in racialized communities. Royal Society of Canada, Ottawa
19. Shingler B (2021) Montreal community groups press, once again, for race-based data on COVID-19. CBC News, 16 Feb 16
20. Fineman M (2014) Vulnerability, resilience, and LGBT youth. Temple Polit Civ Rights Law Rev 23(2):307–330
21. Tully J (2008) Public philosophy in a new key: volume 1, democracy and civic freedom. Cambridge University Press, Cambridge
22. Lee C (2021) Homeless encampment residents, advocates protest park clearings at Toronto rally. The Globe and Mail, 26 July
23. Geuss R (2008) Philosophy and real politics. Princeton University Press, Princeton

Ubaka Ogbogu is a Professor, Associate Dean Research, and Katz Chair in Health Law at the Faculty of Law at the University of Alberta in Edmonton, Canada. Dr. Ogbogu is a recipient of the Confederation of Alberta Faculty Associations Distinguished Academic Early Career Award. He holds a doctorate in law from the University of Toronto, a Master of Laws degree from the University of Alberta and undergraduate degrees in law from the University of Benin, Nigeria and the Nigerian Law School. Ogbogu's scholarly work is focused broadly on the ethical, legal and societal implications of novel and emerging biotechnologies and associated research. His publications have explored a diverse range of issues in this field, including the ethical and legal issues associated with stem cell research, gene and engineered cell therapies, biobanks, germline gene editing and assisted reproductive technologies. As a multidisciplinary scholar, his teaching and research activities explore and cut across various fields, including health law, bioethics, science policy, science and technology studies, public health, legal history and legal philosophy. He has led or been involved in many prominent national and international biotechnology policymaking activities and writes and comments frequently in the popular press on matters relating to the impacts of biotechnology and science on society. Ogbogu has served as the Chair of the University of Alberta's Research Ethics Board No. 2 and as a member of numerous boards and councils, including the Health Quality Council of Alberta, Council of Canadian Academies Expert

Panel on Somatic Gene and Engineered Cell Therapies, the Council of Canadian Academies Expert Panel on Medical Assistance in Dying, the Canadian Institutes of Health Research (CIHR) Stem Cell Oversight Committee, the Canadian Institutes of Health Research Governing Council's Standing Committee on Ethics, and the International Society for Stem Cell Research Task Force on Guidelines for Stem Cell Research and Clinical Translation.

Chapter 14
Vulnerability as a New Perspective on Ethical Challenges in Healthcare

Henk ten Have

And if we hate the virus for the vulnerability it exposes, we ought not for that reason conclude that the absence of the virus will eradicate that vulnerability.
Judith Butler, What world is this? A pandemic phenomenology, 2022

Abstract Vulnerability is a popular notion in recent ethics literature. It is used most often in association with globalization, global health, and pandemics. This contribution examines the use of this notion in ethical discourses concerning health and healthcare. The COVID-19 pandemic illustrates that vulnerability has two interrelated dimensions: a persistent one, reflecting that being human means being vulnerable; every human being may become infected, ill or may even die because of the viral threat; and a variable one, making some humans more vulnerable to the virus and its damaging effect, or as a result of the stringent public health measures that are taken. These two dimensions require that moral debates about healthcare have a more encompassing and differentiated approach beyond the dominating emphasis on personal autonomy and individual responsibility.

Keywords Anthropological vulnerability · Special vulnerability · Global bioethics · Globalisation · Autonomy · Responsibility

1 Introduction

Since the turn of the millennium, the notion of vulnerability is increasingly used in a broad range of scientific disciplines and practical activities. A search in PubMed shows that 'vulnerability' as a general keyword has been used 8.7 more often in

H. ten Have (✉)
Center for Global Health Ethics, Duquesne University, Pittsburgh, USA
e-mail: tenhaveh@duq.edu

Department of Bioethics, Anahuac University Mexico, Mexico City, Mexico

S. Achella and C. Marazia (eds.), *Vulnerabilities*, Integrated Science 18,
https://doi.org/10.1007/978-3-031-39378-5_14

scientific publications in 2022, compared to those in 2000 (Table 1). The overall annual growth rate of publications since the beginning of scientific publishing is estimated at 4.10% [1]. Since 1952 until 2018, the annual growth rate has been 5.08%. Although these estimates are based on publications in the physical and technical sciences, as well as the life sciences (including health sciences), they might suggest that the use of the keyword 'vulnerability' expanded more than the increase in number of publications in general.

This contribution will focus on vulnerability in the context of ethics, particularly bioethics. The keywords of 'vulnerability' and 'ethics' have been used in several thousand publications since 1972, and more frequently over the last twenty years (6.1 times more often in 2022 than in 2000). Combining 'vulnerability' with the specific keyword 'bioethics' produces a lower number of publications. In this area, identifying the keyword 'vulnerability' appears for the first time in 1978. In the specific domain of global bioethics, it is first used in 2002. However, there are two interesting connections worth mentioning. One is the combination 'vulnerability' and 'globalisation,' first used in 1951, and referring to a rapidly growing number of publications in the last two decades (21.4 more in 2022 than in 2000). The other is the combination of 'vulnerability' and 'environment,' first used in 1939, and now used 16.5 times more often to label publications than in 2000 (Table 1).

Since bioethics is the scientific discipline concerned with ethical issues related to medicine, life sciences and associated technologies as applied to human beings, it is helpful to explore how often the notion of vulnerability is employed in connection to health and disease [2]. A PubMed search of the relevant notions shows that vulnerability is for the first time associated with health and disease in the 1940s (Table 2).

Table 1 Vulnerability: general and specific uses (PubMed search November 2022)

Keywords	2022	2000	First publication	>20 annual publications	Total number of publications	Multiplication factor since 2000
Vulnerability	16,809	1938	1898	Since 1968	182,673	8.7
Vulnerability and ethics	472	78	1972	Since 1988	8349	6.1
Vulnerability and bioethics	67	16	1978	Since 1995	1127	4.2
Vulnerability and global bioethics	7	0	2002	–	88	7
Vulnerability and globalization	2336	109	1951	Since 1988	17,781	21.4
Vulnerability and environment	3289	199	1939	Since 1979	31,206	16.5

Relevant publications have multiplied since 2000. Interestingly, the keyword 'vulnerability' is more often associated with health than with disease. The largest growth rate is for publications focused on global health; the number of publications with keywords 'vulnerability' and 'global health' started to increase substantially since 1999, and multiplied 71.1 times between 2000 and 2022.

A more specific search in the PubMed database aims to clarify the connection between the notion of vulnerability and infectious diseases (Table 2). Associations between the keywords 'vulnerability' and 'infections' are made in occasional publications since 1926, in more than 100 annual publications since 1991, in more than 1,000 a year since 2013, while the number of relevant publications increases steeply since 2020. The number of publications associating 'vulnerability' with 'pandemics' was significantly lower. The association is made for the first time in 1983. It is mentioned in more than 20 annual publications since 2006. The majority of this type of publications (69.5%; 746 out of 1073) appears during the past four years (2019–2022). In this same period of time, more than 11,000 publications combines the keywords 'vulnerability' and 'COVID-19'. This is 3.5% of all publications with the keyword 'COVID-19' since 2019 (a total of 309,524 according to a PubMed search with this keyword alone).

From this relatively simple search in the PubMed database it can be concluded that 'vulnerability' is increasingly used to characterize publications as a keyword, especially since 2000. Since the 1970s it is also used in combination with 'ethics' and 'bioethics'. The notion of vulnerability plays a particular role in medicine and health sciences, being associated with 'health' and 'disease', starting in the 1940s and expanding since the turn of the millennium. Since that time, it seems particularly

Table 2 Vulnerability and health or disease (PubMed search November 2022)

Keywords	2022	2000	First publication	>20 annual publications	Total number publications	Multiplication factor since 2000
Vulnerability and health	10,115	597	1949	1976	88,384	16.9
Vulnerability and global health	1422	20	1951	1999	9453	71.1
Vulnerability and disease	5346	687	1941	1975	57,376	7.8
Vulnerability and infections	4030	185	1926	1978	28,239	21.8
Vulnerability and pandemics	295	5	1983	2007	1073	59.0
Vulnerability and COVID-19	3971	0	2019	2020	11,128	3971

fruitful in publications on global health. The notion of vulnerability is furthermore associated with infectious diseases. Its association with infections is substantial and rather old, while publications relating the notion with 'pandemics' is more recent.

2 The Concept of Vulnerability

An analysis of the notion of vulnerability in the bioethical literature shows that it is initially employed in the context of research ethics but expanded into other contexts, especially after the adoption of the *Universal Declaration on Bioethics and Human Rights* in 2005 [3]. The Declaration is the first normative document that stated explicitly that respect for human vulnerability is a fundamental principle of global bioethics. Rather than sensitizing medical researchers that some individuals and groups can be exploited in research and need protection, it is now used to clarify that certain contexts such as socio-economic conditions, poverty, violence, and discrimination make individuals and groups vulnerable. This expanded application of the notion is in fact a reflection of the widening of bioethical discourse itself. While bioethics since its emergence in the 1970s developed rapidly as a new discipline it remained primarily focused on moral challenges faced by patients and physicians in clinical medicine, like medical ethics had traditionally done for centuries. The new discipline, however, changed the power balance in medical interactions from physician paternalism to patient autonomy. The concern with vulnerable subjects in medical research expressed the individualistic orientation of the new bioethics. Around the turn of the millennium, highlighted by the above-mentioned Declaration, a broader approach emerged that provided ethical perspectives beyond the individual concern of clinical medicine and bioethics, taking into account the significant role of the social context and the environment in health and disease. Van Rensselaer Potter, the researcher who coined the term 'bioethics' in 1970 argued that professional medical ethics needed to be replaced by a more encompassing ethical approach which combines knowledge of the life sciences with the wisdom of moral traditions, hence the name 'bioethics'. The primary concern of this new discipline should be the question how humankind can survive, thus locating the moral anxieties of individual patients and doctors in a wider context. However, he was disappointed that the subsequent development of the new field of bioethics continued to reiterate the individual perspective of the traditional medical ethics. He thereupon advanced since 1988 the notion of 'global bioethics' to point out that ethical challenges to healthcare and medicine are increasingly worldwide, relating to economic development, environmental degradation, poverty and deteriorating social-economic living conditions for numerous populations. These challenges require an ethical approach that goes beyond the usual individualistic perspective, and that needs to supplement the concerns with individual autonomy with ethical principles such as solidarity, social responsibility, sustainability, and justice. The emergence of global bioethics since 2000 as demonstrated in the UNESCO Declaration, gave rise to a range of ethical principles to examine and address the moral challenges of contemporary medicine as well as life sciences. Against this background, the notion of vulnerability also expanded its individualistic focus into a wider consideration of the social and environmental context of human existence [4].

Because the notion of vulnerability became increasingly used in a broader setting and in a range of disciplines and practices, disputes about the concept are not uncommon [5]. There is no universal agreement about the concept. The UNESCO Declaration for example provides no definition or description of what vulnerability is. Nonetheless, usage of the concept in the bioethical literature demonstrates two common interpretations of the notion: an ontological one that relates vulnerability to the fact of human existence itself, and a circumstantial one that associates it to the conditions and relationship within individual human beings are living [3]. This distinction between persistent and variable vulnerability was first made by philosopher Onora O'Neill, and since then reiterated by many scholars with often different terminology [6]. However, most scholarly papers define vulnerability as either a persistent or a variable characteristic of human beings [7]. This ignores that in fact both dimensions are relevant to understand the concept of vulnerability. The first dimension explains the anthropological condition: the human condition is characterized by fragility and weakness. Human beings in general are vulnerable, because their bodies are embedded in environments that can harm them. Since being human means being vulnerable, this dimension can best be termed 'anthropological vulnerability'. At the same time, human beings are not isolated entities but embedded in relationships with other living beings and with different types of environments. This embeddedness makes individuals vulnerable, and specific individuals and groups more vulnerable because they are more exposed to threats and the possibility of harm than others. Some people are rendered vulnerable by the social and economic conditions in which they live [8]. This type of vulnerability is best termed 'special vulnerability' [5, p. 124–148].

The implications of anthropological and special vulnerability are different. The first dimension reflects the philosophical point of view that vulnerability is a generalized, inherent and shared characteristic of human beings as embodied agents [9]. The implication is that being vulnerable is not an individual feature but a generalized condition of the human species. It also articulates that being human not merely implies agency but simultaneously susceptibility and passivity; humans are often exposed to forces beyond their control, and do not have full control of their existence. Finally, vulnerability is not a negative qualification of human existence referring to weakness, frailty, and lack of power. On the contrary, it signifies potentiality, openness to change and transformation. Recognizing vulnerability as a shared condition will enable new forms of cooperation, solidarity and community [10]. Social institutions have been created in response to this type of vulnerability [11]. Anthropological vulnerability is furthermore a core notion in international human rights language. Because of their shared vulnerability humans feel pain, and can suffer. They are dependent on others to grow and mature, to become autonomous individuals and to be cared for in illness and ageing. They need social support and legal protection, and have built social and political institutions to provide collective security. Human rights have emerged because human beings share the capacity to recognize pain and suffering in others.

Common vulnerability is therefore the foundation of human rights [12]. Acknowledging anthropological vulnerability as more than an individual characteristic implies an appeal to a different set of ethical values than emphasized in mainstream bioethics: care, solidarity, justice, and international responsibility, going beyond the perspective of individual autonomy, power, protection, and damage control [13, 14].

The dimension of special vulnerability reflects a political rather than philosophical interpretation of the concept. It assumes that vulnerability can be the product of specific circumstances that require remediate action. It is not an abstract notion but manifests itself in everyday life, especially in healthcare settings. Even if human beings share the same inherent and common vulnerability as argued in the philosophical perspective, vulnerability can be exacerbated because of human interconnectedness and living conditions, making some of us more vulnerable. The political perspective emphasizes that vulnerability is made, produced, or generated within specific conditions within which individual human beings happen to exist: poverty, homelessness, discrimination, inequalities in access to healthcare and health insurance, poor socio-economic circumstances, and environmental degradation. These conditions may expose individuals and groups to exploitation, mistreatment, abuse, stigmatisation, and disrespect. They make certain groups particularly vulnerable, such as racial minorities, the economically disadvantaged, people who are disabled, very sick or institutionalized [15].

Special vulnerability is related to processes of globalization that have resulted in a world with more and new threats. At the same time, these processes have undermined traditional protection mechanisms such as social security and welfare systems, and family support mechanisms, thus eroding the abilities of individuals and communities to cope with serious threats. The fact that the world has become increasingly interconnected and interdependent has created a sense of mutual vulnerability. Being vulnerable is often the result of a range of external conditions, and therefore beyond the power and control of individuals. It is argued, for example, that the landscape of medical research has significantly changed [16]. It is now a global enterprise, requiring a broader ethical framework. Globalization has created an asymmetry of power of which vulnerability is one of the major symptoms [17]. It is also indicated that there is growing vulnerability, especially of women in developing countries, related to neo-liberal, global economic policies [18]. Failing states are blamed for increasing vulnerability due to the persistence of poverty and hunger [19]. And it is observed that the discourse of vulnerability has particularly emerged and expanded in the context of global phenomena such as natural disasters and the pandemic of AIDS [20].

The association of vulnerability to globalization requires a broader interpretation of the concept than is usual in mainstream bioethics which used to consider it primarily as impaired or failed autonomy. An influential description of vulnerability in research ethics documents is: "a substantial incapacity to protect one's own interest" [21]. The moral principle of respect for autonomy is the framework within which the notion of vulnerability is interpreted and understood. Vulnerability is primarily regarded as an individual weakness; it indicates that certain individuals cannot protect themselves. For example, in clinical research vulnerable persons either lack decisional capacity or lack adequate information so that they need to be protected against possible exploitation. Free and informed consent can therefore eliminate the vulnerability of potential research subjects. In this perspective, vulnerability essentially is limited autonomy. However, if vulnerability is a global phenomenon and produced by structural social, economic and political determinants that disadvantage people, it is not merely an individual affair. The notion is then more related to the ethical principles of justice, solidarity and equality than individual autonomy. The implication is that bioethics needs to adopt a broader normative framework since the ethical principles that dominated bioethical discourse during the past 50 years are no longer sufficient to provide guidance at the global level [22].

3 COVID-19 and Vulnerability

Vulnerability has become a significant concern during the COVID-19 pandemic (see Table 2). It has been identified as a core principle in many policy statements. Public health measures were often justified with the appeal to protect the most vulnerable citizens. Vaccine distribution schedules usually prioritized vulnerable populations [23]. The pandemic highlighted that some individuals and groups were more affected by the viral disease than others. It was further recognized that health professionals and care institutions themselves are vulnerable [24].

The pandemic clearly illustrates the two dimensions of vulnerability. On the one hand, anthropological vulnerability is revealed in the fact that all human beings where ever they live can be infected. Everybody is a potential patient, and therefore obliged to shelter in order to prevent infection. Human beings are necessarily embedded in natural environments; they cannot separate themselves from their biological surroundings. Micro-organisms such as viruses are essential components of the biosphere and necessary for the sustenance of life. Since humans live in a virosphere, viral infections cannot be eliminated. As key components of the living world many viruses play a positive role in the biosphere. The healthy human body is inhabited by massive numbers of viruses, and viral material is incorporated in our genes [25]. However, the anthropological dimension of vulnerability not only refers to the biological constitution but also to the fact that humans are social beings, i.e. connected to their environment and related to other beings. Vulnerability exists because human beings are open to the world, they engage in relationships with other persons, and interact continuously with the world. This openness and interaction is a positive phenomenon; it is the basis for exchange and reciprocity between human beings. We cannot come into being, flourish and survive if our existence is not connected to the existence and flourishing of others. At the same time, this world-openness is also potentially harmful and damaging, exemplified in the continuous risk of being infected. This risk can be diminished but not completely avoided since humans cannot discard their social nature [26].

On the other hand, COVID-19 has highlighted the dimension of special vulnerability; it has revealed and amplified previously existing vulnerabilities, showing that some human beings are more susceptible to harm than others. The pandemic demonstrates that healthcare workers are more frequently infected, because of their increased exposure. In addition, several other groups are more at risk for different reasons: older people, persons with underlying conditions or compromised immune systems, socio-economically disadvantaged people, indigenous populations, and racial and ethnic minorities. In the U.S., waves of COVID-transmission revealed structural vulnerabilities, first affecting nursing homes and long-term care facilities, then minority populations and immigrants, and next correctional facilities [27]. The experience with the pandemic furthermore saw the resurgence of discriminatory practices such as racism and ageism.

COVID-19 has also created new vulnerabilities since people may become vulnerable due to the policy responses to the pandemic [28]. Lockdowns produced sudden loss of income, livelihood, and food for numerous people, especially in developing countries [29]. They initiated an abrupt disruption of social contacts, for example with visitor restrictions in long-term care facilities and stay-at home orders, exposing many to isolation, loneliness, and depression. Closing of schools deprived children not only from education but also from adequate nourishment in countries with school food programmes. Distribution of vaccines has been criticized as reinforcing global inequity, and making less-resourced countries more vulnerable to serious consequences of infection. Moreover, the priority for hospital care for COVID patients caused cancellation or postponement of interventions and treatment for patients with other diseases, making them more vulnerable to the harm of their disorders, and reducing access to healthcare for many people. At the global level, public health, medical treatment, and preventive programmes were affected or delayed [30]. For example, in 2020 tuberculosis deaths have increased since 2005. The number of people treated for drug-resistant tuberculosis decreased with 15%, and those receiving preventive treatment for tuberculosis infection with 21% [31].

4 Vulnerability After COVID

Although the notion of vulnerability has been recognised and examined since decades, the pandemic made clearly visible that in daily existence people are confronted with vulnerability at multiple levels: as individuals, as persons relating to other beings, as citizens within complex societies and fragile democracies, as consumers of globalized trade, as professionals providing healthcare, as patients within care institutions, as workers in certain enterprises, as inhabitants of low-resource countries [32]. The pervasiveness of vulnerability in pandemic experiences should instigate the development of a broader moral grammar to understand and address the normative challenges of contemporary healthcare. The reason is that these experiences articulate the significance for human existence of liminality, connectedness and community.

Pandemics are liminal events; they disrupt the normal ways of living, acting and thinking, and they call for a transition to a new phase [33]. Old patterns are dissolving but new ones are not established, producing experiences of ambiguity, uncertainty, fear, and disorientation. Liminality indicates that we are living in a borderland where existing frames of reference have become unclear and uncertain. Human beings are always situated, i.e., they find themselves in situations of specific and concrete circumstances, dependent on gender, age, race, character, education and particular circumstances. Some situations, which Karl Jaspers has called 'limit situations' are inescapable: nobody can avoid death, serious illness and suffering. The security of existence disappears and humans are confronted with their vulnerability. In such situations, humans become aware not merely of the limitations but also of the possibilities of existence, and they can go beyond them in communication with others. Such limit situations provide the possibility of a transition to a new orientation of life [34]. There is a strong tendency to deny liminality, for example in the omnipresent use of the war metaphor during the various waves of COVID-19 and the reassurance of policy-makers to return to normal as soon as possible. Nonetheless, pandemic experiences have highlighted human vulnerability; the threat of disease and the possibility of death were no longer abstract events for many people.

The pandemic has reinforced the awareness that connectedness is a basic feature of this era of globalization. Citizens in one country will be exposed to diseases when they emerge in other countries. Closing borders, restricting travel, and concentrating on national interests have had only a limited effect on the dissemination of COVID-19. Vulnerability to infections is not confined to specific individuals, populations and nations but all humans are facing this threat together. Being situated in a web of connections is a precarious experience. Because their bodies position them in the world, human beings are exposed to the world and other persons, necessarily implying vulnerability. If human beings not merely interact with each other but belong together and are mutually dependent, the emphasis on the notion of individual autonomy is no longer sufficient to address and explain the challenges of the pandemic, but common interests, mutual support, social responsibility, cooperation and solidarity should have a significant role in inclusive and comprehensive ethical discourse. Furthermore, the experience of connectedness clarifies our dependency on the planet. Human flourishing and survival crucially depends on the material conditions provided by ecosystems and biodiversity. Without a healthy biodiversity providing water, food, security and medicine, human health is unthinkable. A rapidly increasing number of species becomes extinct because their habitat is irreversibly degraded and destructed. Humans are undermining their own existence and survival, and as the concept of Anthropocene accentuates, their survival as a species is at stake [35]. The issue of uncertain survival not only clarifies that it is impossible to abstract humanity from nature. Humans are part of a biotic community of soil, water, plants, and animals, or more broadly, part of the Earth system which they themselves are jeopardizing. Viral threats precisely illustrate that human health is intrinsically connected to planetary health. They are not natural events but produced by human behaviour, exploiting the planet and destroying biodiversity for the sake of economic growth. Environmental degradation, and the resulting risks of emerging infectious diseases, is associated with an economic world order that proceeds with the assumption that humans and nature are separated, and that nature can be regarded as a resource to be exploited and commodified. That this assumption is completely mistaken is demonstrated in the experiences with the pandemic, showing the connectedness not only between humans but also between humans and nature [26, pp. 115–121].

Finally, pandemic experiences have articulated the importance of community for human flourishing. The notion of community (like related notions of culture, tradition, history and social practices) is usually considered to have secondary relevance from the standpoint of individualism. The COVID pandemic underlines that individuals are not isolated, abstract entities but social beings; they are flourishing within a communal context. Societies are not mere collections of individuals but have their own history and evolution, producing and determining who human beings are and what they will become; their moral status is not so much dependent on their particular individual characteristics but rather on relations and interdependencies with other beings. This point of view is not new; it has been advocated in many critical discourses over time: communitarianism, existentialist philosophy, anthropological medicine, feminist ethics, indigenous ethics, and non-Western philosophies. They focus attention on the fact of dependency and vulnerability, on sociality as a necessary condition for personal identity, and on embeddedness as a precondition for moral agency. In public and policy debates during the COVID pandemic, individual and communal interests are often opposed. Particularly when the ideology of individualism prevails, the human being is regarded as rational self-regarding actor, as *homo economicus* who is motivated by self-interest and is self-managing his or her life driven by calculations to maximize the expected utility for him or herself. In healthcare, this government of life should be encouraged by treating patients as responsible consumers who actively seek information and produce health as the outcome of their choices. What is needed is correct information and proper education because health is primarily a matter of personal responsibility. In many countries, the governments therefore appealed first of all to the individual citizens to show this responsibility in implementing public health measures. The pandemic, however, illustrates that this opposition of individual and community is false since individual behaviour affects the well-being of the community. Widespread use of face masks will protect not only the individual but also other people against possible infection. Testing will identify whether someone is infected, but it is a warning signal that others may be at risk. The aim of vaccination is not only to protect individuals but society as a whole. In a public health emergency, appeals to self-interest cannot be separated from concerns with the interests of others. Individual decisions whether or not to adhere to public health measures have an inherently social dimension. Appeals to individual responsibility will therefore not be sufficient without articulating social responsibility, and without creating the social, political and economic conditions for the exercise of responsible autonomy [36].

5 The Need for a Global Ethics Perspective

The significance of liminality, connectedness and community illustrate the need for a normative framework to understand and address the normative challenges of contemporary healthcare that is broader than the dominating current one. The COVID-19 pandemic exposes vulnerability as a shared and global phenomenon, not only of individual persons but also of populations and subpopulations, and at the same time of the systems that have been built to protect humans against vulnerability, particularly healthcare systems. One of the basic fears during the COVID-19 pandemic was that healthcare institutions might collapse. In most countries they have to learn how to be more resilient and able to cope with the surge of infectious cases, sometimes in the hard way, being unable to provide sufficient protection to care workers, having insufficient testing capacities, triaging patients for intensive care, and descaling care services necessary for non- COVID patients. The pandemic made clear that vulnerability cannot be construed as an individual affair, although it is manifested in individual persons. The notion can no longer be framed, as is usual in mainstream bioethics in terms of the ethical principle of respect for autonomy. In the broader perspective of global bioethics, the view that individual persons are autonomous and in control is challenged. Since the human conditions is inherently fragile, all human beings are sharing the same predicament. As social beings and in order to remediate the vulnerability of existence, humans have developed institutions and social arrangements to protect themselves. This is neither an individual accomplishment nor a threat. Vulnerability means that we are open to the world; that we can engage in relationships with other persons; that we can interact with the world. It is not a deficit but a positive phenomenon; it is the basis for exchange and reciprocity between human beings. The notion of vulnerability therefore refers to interdependency and mutuality, the needs of groups and communities, not just those of individuals. Other ethical principles besides respect for individual autonomy are important, such as justice, solidarity and equality.

In the perspective of global bioethics it is at the same time recognized that special vulnerability is a symptom of the growing precariousness of human existence and is exacerbated in certain conditions. This implies that the social context can no longer be ignored in bioethical analysis and that attention should go to the distribution and allocation of vulnerability at the global level. Instead of focusing on individual features, ethical analysis should examine and criticise the external determinants that expose individuals to possible damage and harm. It also means that individual responses are insufficient; what is needed is a collective response, in other words social and political action. Global bioethics therefore attempts to overcome the ambiguity of vulnerability in contemporary ethical discourse. Over the past few decades, vulnerability has become a relevant and important notion in bioethical debates because processes of globalization have widened the 'space of vulnerability' [19]. Particularly neoliberal policies have increased the precariousness of life across the world. Mechanisms of social protection have declined, and people have diminishing abilities to cope with threats and challenges. Societies have become subservient to the needs of the global economic system. Neoliberal policies are based on the assumption that a human being is self-interested and rational, as well as responsible for his or her own well-being. As *homo economicus* the individual is motivated by minimizing costs and maximizing gain for him or herself. In this perspective, humans relate primarily to others through market exchanges. Citizenship, the public sphere and social networks erode since there are only individuals and commodities that can be traded [37].

In addressing vulnerability, contemporary bioethics is often using the same basic assumption, arguing that vulnerability should be reduced through empowering individual autonomous decision-makers and reinforcing personal responsibility. Abstracting from the social dimension of human existence, and neglecting the damaging impact of market mechanisms on social life, bioethicists contribute to policies and guidelines in the hope to redress the impact of vulnerability. What is a symptom of the negative impact of a one-dimensional view of human beings (and resulting policies) is remedied with policies based on the same type of view. As long as the problematic conditions creating and reinforcing human vulnerability are not properly analysed and criticized, bioethics will only provide palliation.

The paradox is that the ethical discourse of vulnerability has developed in association with increasing processes of globalization. It gives voice to today's experiences that everyday existence is more precarious, that we are exposed to more hazards and threats, and that our capacities to cope have decreased. But as long as the production of vulnerability itself is not critically examined, the roots of the problem will not be addressed. Framing vulnerability as deficit of autonomy presents not only one part of the whole story but it also implies a limited range of options and actions. In this sense, the mainstream ethical interpretation of vulnerability is ideological: it directs theoretical and practical attention away from the circumstances that make subjects vulnerable. The perspective of global bioethics focuses attention to the wider context that produces vulnerability; on what Powers and Faden have called "systematic patterns of disadvantage" [38]. This is only possible in an ethical framework that goes beyond the individual perspective and that includes justice, solidarity, care, and social responsibility. The experiences with the COVID-19 pandemic have also learned that directing attention to the root causes of contemporary problems is not enough. Individual security within a context of emerging infectious diseases can no longer be regarded as "a matter of individual choice" [39, p. 168]. When the major bioethical problems of today are produced by the dominance of neoliberal market ideology, and when precariousness, inequality and exclusion are characteristics of the social order of neoliberal globalization, bioethics should redefine itself as critical global discourse. Focusing attention on the social context of human life will not be enough. Bioethics must argue for a reversal of priorities in policy and society: economic and financial considerations should serve the principles of human dignity and social justice, and no longer be ends in themselves. This implies specific strategies for social inclusion but also institutional support. It will be necessary to demonstrate more vigorous advocacy and activism, supplementing academic enquiry. Social inequalities and conditions that produce vulnerability are not beyond social and political control. It will also require that the voices of the disadvantaged, the deprived and the vulnerable are more often heard within the bioethical discourse, involving vulnerable groups in policy development and implementation. Global vulnerability is furthermore transforming the significance of cooperation. Forging global alliances and new networks of solidarity is the only way to address global threats. An individualistic perspective makes it impossible to address the root causes of vulnerability.

6 Conclusion

Vulnerability reflects on the one hand the precariousness of the human condition and the fragility of the human species, on the other hand the radical changes in contemporary human existence due to processes of globalization. Because every human is vulnerable and there is a constant possibility of harm, human beings need each other and must cooperate. They need institutions such as social networks, protective mechanisms, and human rights to survive and flourish. Vulnerability therefore is not just an individual attribute. Mainstream bioethics construes vulnerability as deficient autonomy. It does not take into account that autonomy itself demands appropriate conditions to arise, to develop and to exercise. Vulnerability therefore is misconstrued as an individual attribute whereas it does not direct attention towards the underlying conditions for human flourishing. Vulnerability is not merely inability or deficiency but most of all ability and opportunity. Experiences with vulnerability during the COVID pandemic show that a critical global ethics discourse is necessary that not only understands the root causes of vulnerability, and that is concerned with vulnerable persons but also intervenes through socio-political and economic measures, recognizing that the major bioethical problems of today are produced by the dominance of neoliberal market ideology. It is not surprising that the language of vulnerability is often used by international and intergovernmental organizations. The devastating effects of neoliberal policies are most visible in the developing world. But nowadays, existential insecurity is everywhere, as is demonstrated by the pandemic. Even in developed countries vulnerability is unequally distributed, and some individuals and groups of persons are disproportionately affected by the virus and the public health measures against it. Reflecting on the experiences of vulnerability should move ethics from the concern with individual well-being towards consideration of the social, cultural, political and economic conditions that are appropriate for human flourishing.

Core Messages

- The significance of vulnerability during the COVID-19 pandemic provides the opportunity to better understand the two dimensions of vulnerability: anthropological and special.
- Vulnerability is not only a philosophical but also political concept since it demands to address its root causes in the conditions of human life.
- Processes of globalization have increased the precariousness of human life across the world, making vulnerability a concept that is not only relevant to exceptional

circumstances or to populations in less resourced countries, but nowadays to all societies.

- Ethical discourse concerning health and disease should be redirected towards a global framework that considers vulnerability no longer as individual weakness or deficiency of individual autonomy but as manifestation of socio-political and economic inequality and structural violence.

References

1. Bornmann L, Haunschild R, Mutz R (2021) Growth rates of modern science: a latent piecewise growth curve approach to model publication numbers from established and new literature databases. Humanit Soc Sci Commun 8:224. https://doi.org/10.1057/s41599-021-00903-w
2. This characterization of bioethics is based on Article 1 of the *UNESCO Universal Declaration on Bioethics and Human Rights*, which does not provide a definition of bioethics but delineates its field of work. See: https://unesdoc.unesco.org/ark:/48223/pf0000146180
3. Mergen M, Akpinar A (2021) Vulnerability: an integrative bioethics review and a proposed taxonomy. Nurs Ethics 28(5):750–765
4. Ten Have H (2016) Global bioethics. An introduction. Routledge, London and New York
5. Ten Have H (2016) Vulnerability. Challenging bioethics. Routledge, London and New York
6. O'Neill O (1996) Towards justice and virtue. Cambridge University Press, Cambridge
7. Sanchini V, Sala R, Gastmans C (2022) The concept of vulnerability in aged care: a systematic review of argument-based ethics literature. BMC Med Ethics 23:84. https://doi.org/10.1186/s12910-022-00819-3
8. Victor E, Luna F, Guidry-Grimes L, Reiheld A (2022) Vulnerability in practice: peeling back the layers, avoiding triggers, and preventing cascading effects. Bioethics 36(5):587–596
9. Rendtorff JD (2002) Basic ethical principles in European bioethics and biolaw: autonomy, dignity, integrity and vulnerability—towards a foundation of bioethics and biolaw. Med Health Care Philos 5(3):235–244
10. Solbakk JH (2011) Vulnerability: a futile or useful principle in healthcare ethics? In: Chadwick R, Ten Have H, Meslin EM (eds) The SAGE handbook of health care ethics: core and emerging issues. Sage, London, pp 228–238
11. Fineman MA (2010) The vulnerable subject and the responsive state. Emory Law J 60:251–275
12. Turner BS (2006) Vulnerability and human rights. Penn State University Press, University Park
13. Ten Have H (2015) Respect for human vulnerability: The emergence of a new principle in bioethics. J Bioethical Inquiry 12(3):395–408
14. Delgado J (2021) Vulnerability as a key concept in relational patient-centered professionalism. Med Health Care Philos 24(2):155–172
15. Zagorac I (2016) How should we treat the vulnerable? Qualitative study of authoritative ethics documents. J Health Care Poor Underserved 27(4):1656–1673
16. Eckenwiler L, Ellis C, Feinholz D, Schonfeld T (2008) Hopes for Helsinki: reconsidering 'vulnerability.' J Med Ethics 34(10):765–766
17. Justo L (2004) Participatory research: a way to reduce vulnerability. Am J Bioeth 4(3):67–68
18. Jaggar AM (2002) Vulnerable women and neo-liberal globalization: debt burdens undermine women's health in the global South. Theoret Med 23(6):425–440
19. Watts MJ, Bohle HG (1993) The space of vulnerability: The causal structure of hunger and famine. Prog Hum Geogr 93:43–67
20. Delor F, Hubert M (2000) Revisiting the concept of 'vulnerability.' Soc Sci Med 50:1557–1570
21. CIOMS (1993) International guidelines for biomedical research involving human subjects. CIOMS, Geneva: 10
22. Ten Have H (2014) Vulnerability as the antidote to neoliberalism in bioethics. Revista Redbioetica/UNESCO 5 (1); no 9:87–92

23. Maeckelberghe E (2021) Ethical implications of Covid-19 vulnerabilities in a global perspective. Eur J Public Health 31(suppl 4):iv50–iv53
24. Blauwet CA, Brashler R, Kirschner KL, Mukherjee D (2020) Vulnerability, interdependence, and trust in the COVID-19 pandemic. PM R 12(10):1038–1044
25. Ryan F (2020) Virusphere. Ebola, AIDS, influenza and the hidden world of the virus. William Collins, New York
26. Ten Have H (2022) The covid-19 pandemic and global bioethics. Springer/Nature Publishers, Cham
27. Solis J, Franco-Paredes C, Henao-Martinez AF, Krsak M, Zimmer SM (2020) Structural vulnerability in the U.S. revealed in three waves of Covid-19. Am J Trop Med Hygiene 103(1):25–27
28. Editorial (2020) Redefining vulnerability in the era of Covid-19. Lancet 395(10230):1089
29. Bottan N, Hoffmann B, Vera-Cossio DA (2020) The unequal impact of the coronavirus pandemic: evidence from seventeen developing countries. Inter-American Development Bank. https://publications.iadb.org/publications/english/document/The-Unequal-Impact-of-the-Coronavirus-Pandemic-Evidence-from-Seventeen-Developing-Countries.pdf
30. Meyerowitz-Katz G, Bhatt S, Ratmann O et al (2021) Is the cure really worse than the disease? The health impacts of lockdowns during COVID-19. BMJ Glob Health 6:e006653
31. Pai M, Kasaeva T, Swaminathan S (2022) Covid-19 devastating effect on tuberculosis care—a path to recovery. N Engl J Med 386:1490–1493
32. Flood CM, MacDonnell V, Philpott J, Thériault S, Venkatapuram S (2020) Vulnerable. The law, policy and ethics of Covid-19. University of Ottawa Press, Ottawa
33. Bedyński W (2020) Liminality: black death 700 years later. What lessons are for us from the medieval pandemic? Soc Register 4(3):129–144
34. Thornhill C, Miron R (2020) Karl Jaspers. In: Zalta EN (ed) The Stanford encyclopedia of philosophy. https://plato.stanford.edu/archives/spr2020/entries/jaspers/
35. Ten Have H (2019) Wounded planet. How declining biodiversity endangers health and how bioethics can help. Johns Hopkins University Press, Baltimore
36. Macer D (2020) The foundation and functioning of the world emergency Covid-19 pandemic ethics committee. In: Woesler M, Sass H-M (eds) Medicine and ethics in times of corona. LIT Verlag, Zürich, pp 115–125
37. Kirby P (2006) Vulnerability and violence. The impact of globalization. Pluto Press, London/Ann Arbor
38. Powers M, Faden R (2006) Social justice. The moral foundations of public health and health policy. Oxford University Press, Oxford/New York, p 71–79
39. Harvey D (2005) A brief history of neoliberalism. Oxford University Press, Oxford/New York

Henk ten Have studied medicine and philosophy at Leiden University (the Netherlands) (MD 1976; Ph.D. 1983). He worked as a researcher in the Pathology Laboratory, University of Leiden; as a practicing physician in the Municipal Health Services of Rotterdam before being appointed as Professor of Philosophy in the Faculty of Medicine and Faculty of Health Sciences, University of Limburg, Maastricht; and subsequently as Professor of Medical Ethics and Director of the Department of Ethics, Philosophy and History of Medicine in the University Medical Centre Nijmegen (the Netherlands). In 2003 he joined UNESCO as Director of the Division of Ethics of Science and Technology. From 2010 until 2019 he was Director of the Center for Healthcare Ethics at Duquesne University in Pittsburgh (United States). In July 2019 he retired and was made Emeritus Professor at Duquesne University. His latest books are *Global Bioethics: An Introduction* (2016), *Vulnerability: Challenging Bioethics* (2016), and *Wounded Planet: How Declining Biodiversity Endangers Health and How Bioethics Can Help* (2019). He is editor of the *Encyclopedia of Global Bioethics* (3 volumes, 2016) and (with Bert Gordijn) editor of the *Handbook of Global Bioethics* (4 volumes, 2014). Since 2021 he is Research Professor at the Department of Bioethics, Anahuac University, Mexico.

Chapter 15
Vulnerability, Interest Convergence, and the COVID-19 Pandemic: Lessons from the Future

Jane Cooper, Zamina Mithani, and J. Wesley Boyd

To build community requires vigilant awareness of the work we must continually do to undermine all the socialization that leads us to behave in ways that perpetuate domination.
— bell hooks, Teaching Community: A Pedagogy of Hope

Abstract In this chapter, we will explore how COVID-19 has made us collectively vulnerable to illness and death, although marginalized and minority communities have been particularly hard hit. In stark fashion, the pandemic has shown us that as a discipline, bioethics can no longer engage in business as usual but instead needs to be reimagined. We explore two conceptions that will help explore how this collective vulnerability has occurred with COVID-19, and how the change that occurs often is not sustainable or truly "ethical". The first is Rawl's "veil of ignorance" and the second is Derrick Bell's "interest convergence" theory, which posits that at some points in a movement, things are only good because the party with power has an agenda that aligns or "converges" with the non-dominant group, but not because there is actual moral agreement. Although these theories are largely descriptive, Heckler and Mackey's novel theory on interest cognizance suggests that there are normative ways in which a movement becomes sustainable because there is not just superficial "buy-in" but actual moral agreement. From that analysis, we posit a

J. Cooper (✉)
University of Toronto Faculty of Law, Toronto, Canada
e-mail: jane.falliscooper@mail.utoronto.ca

Z. Mithani
University of British Columbia Faculty of Medicine, Vancouver, Canada
e-mail: zaminam@student.ubc.ca

J. Wesley Boyd
Baylor College of Medicine, Houston, USA

Department of Global Health and Social Medicine, Harvard Medical School, Boston, USA

J. Wesley Boyd
e-mail: jwesley.boyd@bcm.edu

S. Achella and C. Marazia (eds.), *Vulnerabilities*, Integrated Science 18,
https://doi.org/10.1007/978-3-031-39378-5_15

path forward for bioethics that goes beyond traditional principles and incorporates novel considerations such as connectivity. Given global warming and other looming catastrophes, including possible future pandemics, such a reimagining is necessary in order to move forward in the face of extreme global vulnerability.

Keywords Vulnerability · Ethics · Justice · Racism · Ableism · Pandemic · COVID-19 · Bioethics

1 Introduction

The Covid-19 pandemic has been—and continues to be—a time period of great historical significance in modern human history. The past few years have helped to make the successes and failures of the current global order and cooperation clearer, and have provided a look at global cooperation in the face of a collective threat. We have seen courageous examples of human ingenuity, and communities rallying together to support their most vulnerable. We have also seen vaccination inequality, racism, rampant misinformation, and troubling apathy for fellow human beings emerge as concerning trends as responses to the Covid-19 crises.

In this chapter, we explore how COVID-19 has made us collectively vulnerable to illness and death, although marginalized and minority communities have been particularly hard hit. The pandemic has shown us that as a discipline, bioethics can no longer engage in business as usual, and instead needs to be reimagined. We engage with two conceptions which illustrate how our collective vulnerability led to temporary positive change: Rawl's "veil of ignorance" and Derrick Bell Jr.'s "interest convergence". We then discuss Heckler and Mackey's novel interest cognizance theory as both an important way of conceptualizing our failure to make long standing equitable changes, including during the COVID-19 pandemic, and as hope for the future. Building on this analysis, we discuss a path forward for a bioethics that moves beyond traditional principles and incorporates novel considerations such as connectivity. These discussions are also placed in the context of global warming and other looming catastrophes of human vulnerability, that illustrate the necessity of such a reimagining.

2 Temporary Vulnerability from COVID-19 Can Bring Some Good

While there are many ongoing societal challenges that have been exposed, created, or exacerbated by this pandemic, the near universal quarantine implemented in the early stages of the pandemic seemed to create a momentary sense of population-level vulnerability. At that time all of us appeared to face the risk of infection and death, irrespective of our race or social status, among other characteristics. Thus COVID-19 had, for a brief moment, created a near universal sense of vulnerability. But only for a moment. As time passed, COVID-19 brought out massive inequities in global access to healthcare and vaccinations, and country-specific problems of equity within how members of individual nations were differently affected by and treated for this virus. The disparities faced by Black, Latine, Indigenous, disabled, incarcerated and undocumented individuals, and many other minorities in the United States with respect to health outcomes and death became increasingly apparent as the pandemic wore on.

Despite the vulnerabilities that COVID-19 created and continues to exacerbate, some of the adaptations that COVID-19 forced in terms of quarantining and isolating brought out slivers of good. For example, although COVID-19 dramatically limited the way that in-person events have traditionally been held, forcing events and communication to be online and virtual has allowed much wider access to events for many individuals. This expanded access benefitted groups that had been historically locked out from many in-person events, including those in rural communities, those with familial or other obligations, and perhaps most notably, many people with disabilities. Universities and colleges that previously refused access to recorded lectures and remote attendance were livestreaming lectures and providing virtual content. This move to flexible work-from-home policies "present[ed] employment opportunities for disabled people that they may not have had in the past, even with the Americans with Disabilities Act (ADA)" [1]. Conferences in locations that previously required large amounts of time, money and energy to attend became open to interested parties around the world through Zoom and similar platforms. In other words, places that had historically high barriers to access for people with disabilities prior to COVID-19 suddenly provided access when their mainstream, able-bodied populations were affected due to COVID-19. These changes, such as using online learning and integrating it more seamlessly into classrooms, have changed the ability of those without previous means to attend events in-person and to learn in prominent institutions [2]. Of course, online learning and virtual attendance still has many constraints and can be problematic for some, such as those with intellectual disabilities who might benefit from more in-person time [3]. Nonetheless, the adaptation to quarantine-learning greatly expanded access for many [2].

Additionally, COVID-19 brought with it some expansions of access to free health services such as testing and vaccination, even in jurisdictions where healthcare is privatized such as the United States. More broadly, COVID-19 has led to greater conversations about the need for universal healthcare as "pandemic preparedness," given its importance in both saving lives and reducing costs associated with healthcare [4]. This includes intranational efforts, as well as international efforts tied to the United Nations Sustainable Development Goals [5]. Furthermore, many public health campaigns emphasized a collective approach which sought to unite communities. In Ontario, Canada, for example, the phrase "flatten the curve" encouraged individuals to take proactive group measures, including large-scale testing, quarantining, and social distancing. Other campaigns included messages such as "we're all in this together" and "stronger together" [6, 7]. These campaigns worked to create a sense of togetherness and commonality in some places that resisted past norms that emphasized individuality.

And lastly, while COVID-19 highlighted the brutally unjustice of the American healthcare system, it also brought with it an increased awareness of racial injustice, including in the bioethics community. This corresponded with a reckoning with police brutality and anti-Black racism, in the wake of George Floyd's murder. The unprecedented support that the Black Lives Matter (BLM) movement received, however, may have also been influenced by an increase in *mortality salience*—an awareness that our own death is inevitable [8]. Indeed, death was more tangible during the early pandemic, when death tolls and panic was at its highest, economies were volatile, and quarantine was isolating. Part of the way we can compensate for and address this existential dread is to engage in activities that give us a sense of purpose. Thus, many have posited that there was a synergistic effect between COVID-19 and BLM, which helps make sense of why the BLM movement peaked the way it did during COVID-19, despite many prior decades of senseless Black deaths due to police brutality [8, 9].

3 Regression from Meaningful Change

Unfortunately, many of the positive changes that the COVID-19 pandemic brought—increased access to virtual opportunities, healthcare, and awareness about issues of social justice—have receded over time. A near universal sense of togetherness has given way to numerous instances of malignant individualism, refusal to cooperate with public health measures, and even the exacerbation of social inequities. And as the pandemic has worn on, online accommodations have been dramatically walked back in many academic spaces, effectively removing the temporary good that came from the constraints of the pandemic [10]. As such, the interests of the able-bodied majority, who often prefer the traditional in-person class structure for many good reasons such as enhanced interaction and socialization, have taken priority over any benefit that might arise from a hybrid in-person/virtual arrangement for classrooms. And despite some minor advances due to COVID-19, disparities in both the provision of healthcare in the United States as well as health outcomes not only persist but have increased for minorities and the poor, who have died at disproportionately high rates since the start of the pandemic.

In contrast to those who are poor or who are minorities, there is another group who are highly vulnerable to COVID-19 entirely due to choices they have consciously made, namely those who have flouted medical and scientific directives regarding COVID-19. This group includes individuals who have refused to wear masks or maintain social distance and those who have chosen to forgo getting vaccinated, despite widespread availability of vaccines. Thus, although we are all vulnerable to COVID-19 to some extent, and many of us are vulnerable to a great extent through no fault of our own, another group has actively chosen to make themselves even more vulnerable than they would have been otherwise. We can question the extent to which such choices are freely made, since the amount of misinformation and propaganda that many are exposed to is truly breathtaking.

Despite the fact that the US healthcare system has adapted to the pandemic with more rapid testing as well as expanded vaccine access and research, in reality the overall structure of the American healthcare system is essentially unchanged, despite the incredible burden on emergency departments, intensive care units, and safety-net hospitals brought on by the pandemic.

Additionally, the lack of change in healthcare reflects the lack of change in every institution. In the example of BLM and other movements for racial justice, many institutions promised substantive change to their racist internal structures. Despite their promises for top to bottom institutional change, many institutions' sole response to BLM was to create Diversity, Equity, and Inclusion (DEI) committees and initiatives. While DEI efforts are an important start, there has not been the sweeping change that is truly necessary to dismantle the structural racism that pervades nearly every powerful institution in the US. Thus, although it may have been in many institutions' best interests to try to enact progressive changes in response to the increased social pressure, what we have seen is that once equity issues begin to fade from the media spotlight, the resolve to create real change wanes [11].

Despite this sad reality, the need for real change is nonetheless overdue, because Black people—mostly men—continue to be gunned down by police and because the very structure of our society continues to harm and kill people. For example, one study by Xie and colleagues found that the burdens of an individual's COVID-19 disease varied by demographic groups (age, race, and sex) but were consistently higher in people with poorer baseline health and in those with more severe acute infection. In this study, a nuanced picture of disease burden shows higher levels among Black individuals (e.g., new onset diabetes mellitus, chest pain, substance abuse, thromboembolism, headache, and tachycardia) and females (e.g., chest pain, arrhythmia, headache, smell problems, hair loss, and skin rash) [12]. Concerningly, Black patients are frequently underrepresented in research and intervention development for medical conditions like heart disease and prostate cancer, and may be as well in long-COVID research [13, 14].

Additionally, this lack of research translates to poorer clinical care. For example, in the case of long-COVID care, one Black patient described the lack of adequate resources for her long-COVID symptoms of pain, fatigue and dyspnoea [15], partially because these effects in marginalized communities have been understudied. She says, "*When I pleaded my case to the short-term disability company ... I tried to advocate and explain to them about my symptoms and what I was experiencing and that I wasn't well enough to go to work. And they denied my claim, every time, even still to this day." // They're telling me now that there isn't anything that shows post-COVID affects me from doing my job*" (P13). Thus, the lack of research and clinical care that continues in communities that already experienced the worst health outcomes before the pandemic will likely continue to be disproportionately harmed for years or even decades after the pandemic.

The convergence of the pandemic and BLM exposed, for many, the shortcomings of business as usual within bioethics. In the face of such injustice, bioethicists were now being called upon to not just provide analyses of various situations and encounters, but to actually actively speak out on behalf of justice and against injustice. Bioethics would have to be reimagined, but exactly what would that look like? In order to answer this question, we first turn to the thinking of John Rawls, and then to Derrick Bell Jr.

4 Theories: Veils of Ignorance and Interest Convergence

There have been many explanations for the initial sense of collective vulnerability to COVID-19—even if it was mostly an illusion in the earliest days of the pandemic—followed by a subsequent regression. Some scholars have proposed that COVID-19 has been a "natural version of the Rawlsian thought experiment" [16]. As a brief explanation, John Rawls's veil of ignorance is a thought experiment articulated in his classic text *A Theory of Justice*. In this thought experiment, he puts all persons into his "original position" behind the veil of ignorance, in a state where no one knows any of their characteristics, such as class, race, or gender. From this position, he posits that decision makers, when tasked with designing rules of society, will design a social contract that (1) ensures maximum liberty, and (2) guarantees everyone equal opportunity [17].

In times of COVID-19, while the veil of ignorance was far from perfectly replicated (race, class, gender and other attributes were known but overall vulnerability to COVID-19 was not), overall vulnerability was far more unpredictable than usual. As David Napier explains,

> Because a crisis exaggerates inequalities, it often also pushes previously less vulnerable groups across capability and opportunity thresholds, creating unexpected, new vulnerabilities. With Covid-19, people in service industries without benefits are being hit very hard – especially if their children have been sent home from school. So are people who live in old New York apartment buildings – even posh ones – with shared heating and ventilation systems, especially if they can't open windows. [...] Crises, in other words, give rise to new kinds of vulnerabilities that take us by surprise, creating also new sensitivities [...]. [18]

As COVID-19 became less unpredictable, and individuals returned to the knowledge of their own positionalities, the veil lifted. Those with more privilege generally had greater protections against illness and death (for instance, able-bodied people had a statistically far lower chance of death and or long-term injury; those who had access to high quality healthcare were similarly more likely to recover). Additionally, race-based statistics have shown stark disparities in death rates from COVID-19 in the United States, with Latine, Black, and Indigenous Americans' age-adjusted mortality consistently higher than Asian and White Americans [19]. These COVID-19 disparities only become more stark with each passing day of the pandemic.

The move towards a sense of collective vulnerability and with it a greater drive toward inclusivity and justice, followed by a rapid retreat, also maps on to a different theory: Derrick Bell Jr.'s "interest convergence". To be brief, this theory is central to the field of Critical Race Theory, and was created in light of the seminal US Supreme Court case of *Brown v. Board of Education* that struck down school segregation in 1954. Bell argues that the Supreme Court judges did not rule in favor of equality due to goodwill, or a newfound understanding of equality or anti-racism [20]. Instead, Bell suggests that the court realized that the white ruling class in the US needed to embody and embrace moral leadership in the post-World War II, Cold War era, and that this interest on the part of the court converged with the interests of Black Americans who sought desegregation. More broadly, the theory explains situations in which the interests of those in power temporarily "converge" with the interests of an oppressed group for secondary gains (e.g. political power, moral legitimacy, economic reasons, etc.), leading to the appearance of deep agreement about the issue on a moral, ethical level. In reality, however, there is no actual moral or ethical underpinning for the alignment of interests and the change that occurred does not last. This was true in the case of school segregation—although there might have been interest convergence in the 1950s regarding the need for desegregation, aggressive measures to enforce desegregation were fairly promptly rolled back, and segregation has persisted to the present day. Indeed, almost seventy years later, more than one third of American students attended a predominantly same-race/ethnicity school [21].

Multiple scholars have applied Bell's interest convergence theory to COVID-19. Notably, Heckler and Mackey theorize that the conditions for interest convergence occurred in the "unmasking" or laying bare of systemic racism in public administration in the United States during COVID-19. As an example of this unmasking, Heckler and Mackey point to the higher mortality rates of Black and Latine Americans, and decreases in life expectancy due to the impact of "administrative racism" [22, p. 364].

Heckler and Mackey see administrative racism as context for apparent racial gains made by the BLM movement. In their paper, they argue that "most of white U.S. watched this COVID-19 administrative racism with concern such that the videos of George Floyd's brutal murder at the hands of a police officer created an interest convergence moment" [22, p. 366].

As another example, Meg Peters connects the increase in virtual accommodations for students during COVID-19 to the interest convergence of able-bodied people with people with disabilities [2]. In this application of interest convergence, it is the need of the able-bodied majority to engage with their studies, as well as the institutional need to continue offering courses, that leads to the increase in virtual accommodations.

Both of these examples (BLM and increased virtual learning) have completed the "cycle" that Bell described. As such, anti-Black racism in the United States persists while many "defunded" police departments now have increased budgets, and virtual accommodations have all but disappeared in academic spaces as the interest convergence moment has passed [2, 23].

5 Combination of Theories: Introducing Interest Cognizance

If the veil has lifted, and the interests of the elite decision-making class are no longer convergent with the interests of minorities and other marginalized groups, where does this leave us? Both the veil of ignorance, and interest convergence, are descriptive theories. However, in Heckler and Mackey's 2022 paper "COVID-19's Influence on Black Lives Matter: How Interest Convergence Explains the 2020 Call for Equality and What That Means for Administrative Racism," they lay out a novel alternative ending to the interest convergence problem that they title "interest cognizance" (Fig. 1).

In their explanation of this model, the left cycle represents the aforementioned phenomenon of interest convergence. In the author's example, white urgency creates racial justice advocacy, but as "[t]he most economical way to cultivate moral authority is not to institute anti-racism, but to cultivate the perception of morality using moral licensing and false empathy," hierarchical solutions are proposed that do not truly solve the problems they are purported to address [22]. Moral licensing is the concept that we are more likely to make morally poor choices after making moral ones for a long time, because we think we are already a "moral" person overall. Cascio and Plant in their review article describe moral licensing as the feeling that we have "moral credits", a moral currency, when we do something moral that can be spent at a later date by engaging in immoral behaviour [24]. This leads to a misalignment, and subsequent interest divergence, as the white majority no longer prioritizes anti-racism, and conditions do not substantially improve for people of color.

However, in Heckler and Mackey's model, there is an important alternative. On the right-hand side of the figure, change is "not performative, but substantive" [22, p. 370]. On this side of the model, cultural humility, which requires three proficiencies ((1) marking power, (2) embedding in community and (3) life-long learning).

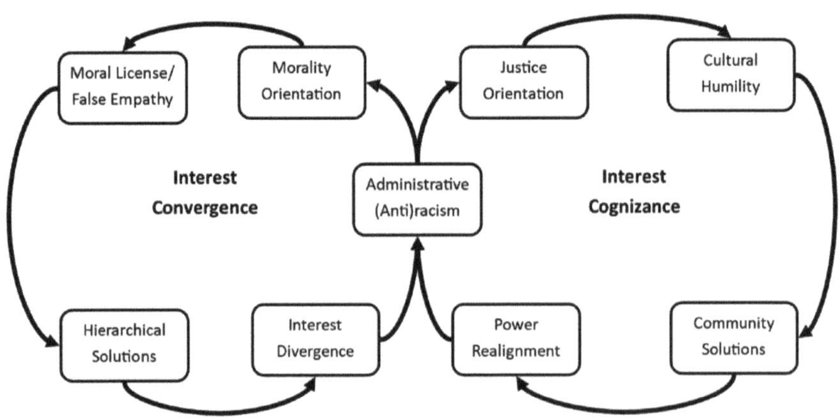

Fig. 1 Interest convergence and cognizance systems [22]

Heckler and Mackey define "marking power" as the extent to which the funders of equity-diversity initiatives are morally separated from the goals of the racial justice movement because of their financial advantage. It calls for more reciprocity between people in power and the community partnerships they make. In this way, change is meaningful, because it is not only decisions that change, but the *very power structures themselves* also change. This prevents, or at least reduces, the likelihood of a subsequent interest divergence. In their example, organizations that employed cultural humility (with attention to the three proficiencies mentioned above), and committed to a restructuring of power (i.e. giving communities and organizations the resources to make decisions that serve their interests), are able to move to an antiracist cognizance system [22].

The idea of interest cognizance is also congruent with the proposal that bioethics should embrace a universal value of vulnerability moving forward, in order to not restrict the good that COVID-19 has illuminated (i.e. access, health, and justice-related examples, as previously discussed) as a moment and not a movement. Ries and Thomson use dementia as an example to show that when we say "vulnerability is a deficit", this can have very unethical consequences for a population by excluding them from research participation. Normatively, we can use this idea to argue that by keeping "vulnerability" as a temporal value that was only useful at the height of COVID-19 when no one knew when quarantine would end and cures would come, can create an environment where we feel that the "vulnerable" only deserve our ethical attention in times of crisis [25], p. 308]. Instead, if we continue to embrace being in a state of perpetual vulnerability, according to the theories of universal vulnerability where it is the "primal human condition," it brings us closer to a place where we take less time to label the vulnerable/marginalized/oppressed as the lesser-than that need our "help", but as a state anyone can and does experience [25, p. 302]. This can be a way that we can go from interest convergence to interest cognizance, as the sample statements in Table 1 highlight.

6 Solutions: Moving Forward

As this chapter aims to demonstrate, we have much to glean from our experiences during COVID-19. With real reflection and humility, we hope that lessons learned can lead to support for efforts to protect both the health of people and the planet. This must include a recognition of human vulnerability to, and interconnectedness with, the rest of the natural world. Just because the acuity of COVID-19 has diminished does not mean that it cannot return in full force, or that new variants and/or new viruses—such as monkeypox—will not arise. In this way, we must challenge the way we think of health as a "stable" right, and instead reveal the truth of health and science as unstable and constantly, dynamically evolving as both the microbial and political world change around us.

Table 1 Sample situations demonstrating interest convergence versus cognizance

Issue	Interest convergence	Interest cognizance
Online learning	Interests of students with disabilities' access to learning converge with the able-bodied student majority as COVID-19 prevents in person learning. As the pandemic recedes, interests diverge.	Access to flexible modes of learning creates the potential for new ways of belonging that reflect changing online and offline communication styles in learners, within and beyond COVID-19.
Vaccine access	Interests of a local population (Global South) in having access to vaccines converge with policy makers (Global North) maximizing world-wide immunity and vaccine sales.	Everyone should have the opportunity to access vaccines in order to build individual and local immunity, as well as to fulfil a global moral obligation for global immunity to avoid future virus variants, this opportunity not discriminate based on economic ability or North/South location.
Anti-racism efforts in bioethics	Interests of marginalized groups converge with the interest of the white majority in bioethics to appear sensitive to BLM for funding and political legitimacy. As news coverage of BLM recedes, interests diverge.	Marginality constantly evolves and bioethics works to continuously do better by being embedded in the communities it addresses through its local structures (e.g. clinical ethics teams, researchers).

Applying the example of accessibility in academia, interest cognizance may happen when we stop conceiving of accessibility as an "extra resource or alternative format in order to correct a deficit" [2]. Instead, if we reimagine access as an experience and process, we can start creating our new baseline of normal and work towards a form of interest cognizance, where we acknowledge that a hybrid solution for education might be the one that creates new opportunities for those individuals with learning disabilities and in-person constraints, and collective opportunities as a society from the innovation and potential that those students can bring to our whole community.

As for the increase in focus on racism and equity issues, the real test comes after media attention dies down, when institutions may complete the interest convergence cycle by either watering down those changes, or fail to adequately fund and enact them, as attention wanes. On the other hand, following interest cognizance an institution that engages in true reform (including power realignment, true community-based solutions, and humility) is able to preserve the changes. Heckler and Mackey's statement that "[e]mbedding in community supports co-created racially just outcomes by emphasizing the value and expertise of all partners," is especially apt for bioethics, which has been long criticized for existing primarily in an "ivory tower" [22]. A bioethics that embraces interest cognizance does not work "for" communities as a temporary consulting service, but is dedicated to embedding itself within community.

In our 2021 paper in the American Journal of Bioethics, we explore "'counter storytelling', an activity which specifically draws on stories that are generally neglected or overlooked by mainstream representations, with the express purpose of highlighting the experiences and conditions of marginalized groups in society" [26, p. 15]. Counter narratives or storytelling causes us to always ask *who else* should be a voice in a dilemma. We continue in our article to conclude that "in bioethics, intentionally incorporating such counter storytelling into ethical dilemmas or issues would allow us to discuss the diverse experiences of those whose voices have historically not been heard within and beyond the four-principle framework. As such, the way we understand, teach, and write within and about bioethics should involve an audit of what is not included within current ethical discourse" [26, p. 16]. Centering the counter narrative can help us do the work of interest cognizance because we are not "othering," but instead are working towards a society where we include these stories in our panels, structures and decision-making bodies. Although we might not be able to include every story in the ways that we engage in ethical engagement, their demographic, perspectives and representation ought to approximate the community they serve.

In bioethics more broadly, this likely means that we ought to push back against hierarchical, "top-down" decision making. The reimagined bioethics that we are considering also ought to move beyond the traditional principled approach of autonomy, beneficence, non-maleficence, and justice and including novel considerations, such as vulnerability, connectedness, community, solidarity and cooperation. A re-envisioned bioethics would draw on concepts from global bioethics and come to fully appreciate the fact that humans are "integrated wholes of body and soul […] embedded within communities [… that] exist in a web of relationships with other beings and the environing world" [27, p. 246]. Understanding this has implications for important global decisions moving forward, especially in the wake of COVID-19, in which wealthier countries manufactured and/or secured the manufacturing of vaccines for their domestic use, without adequate concern with vaccine equity and global vaccination programs. However, "equitable access to COVID-19 vaccines makes a life-saving difference to all countries", as mutations in un/under vaccinated regions inevitably travel, and lead to breakthrough cases in the vaccinated population [28]. This fact, coupled with many wealthy countries stockpiling vaccines via private agreements with pharmaceutical companies—so called "vaccine nationalism"—created a shortage in many poorer countries, especially in the Global South [29]. Developing ethical frameworks that include vulnerability and connectedness among other concepts may aid in creating international policies to prevent the kind of hoarding and stockpiling behaviour with COVID-19 vaccines.

7 Conclusion

Living through the COVID-19 pandemic has changed our relationship with all aspects of life, from work to school to health. It has also changed the public's relationship with ethics, as discussions pertaining to issues of justice, resource allocation, and triage are now as routinely encountered in mainstream media as they are in bioethics forums [30]. As such, the focus of ethics is adapting to ensure that it not only addresses the theoretical components of ethical work, but does justice to a broader range of applications that aim not just to solve present problems, but also to prepare us for and hopefully prevent future problems. As bioethicists, we feel that it is important for our field to engage meaningfully with all aspects of the COVID-19 pandemic, in order to make such prospective suggestions. For us, this includes analysing the applicability of theories such as Rawls' veil of ignorance, and Bell's interest convergence, as well as highlighting the hope and insights that normative theories such as interest cognizance hold. Out of the darkness of this pandemic, we hope that our analysis will help illuminate the promise of a more hopeful future.

Core Messages

- As a discipline, bioethics can no longer engage in business as usual. Instead, it needs to be reimagined in a way that centres equity.
- John Rawls' "veil of ignorance" and Derrick Bell's "interest convergence" theory both provide interesting hypotheses that may explain why some COVID-era equity-centered changes in healthcare are superficial, and not truly "ethical."
- Heckler and Mackey's novel theory on interest cognizance provides a framework for bioethics that goes beyond traditional principles, to incorporate novel considerations such as connectivity. These novel considerations are especially important in the face of global climate change, and other new threats.

References

1. Kanter AS (2022) Our new remote workplace culture creates opportunities for disabled employees. Bill of Health, March 10. https://blog.petrieflom.law.harvard.edu/2022/03/10/remote-work-disability-ada/
2. Peters M (2022) Caring classrooms in crisis: COVID-19, interest convergence, and universal design for learning. Disabil Stud Q 42(1)
3. Natanson H, Strauss V, Frey K (2021) How America failed students with disabilities during the pandemic. Washington Post, May 20. https://www.washingtonpost.com/education/2021/05/20/students-disabilities-virtual-learning-failure/
4. Galvani AP, Parpia AS, Pandey A et al (2022) Universal healthcare as pandemic preparedness: the lives and costs that could have been saved during the COVID-19 pandemic. Proc Natl Acad Sci 119(25):e2200536119
5. Verguet S, Hailu A, Eregata GT, Memirie ST, Johansson KA, Norheim OF (2021) Toward universal health coverage in the post-COVID-19 era. Nat Med 27(3):380–387

6. Pan American Health Organization (2020) "Stronger Together" campaign to highlight importance of psychosocial support during times of stress and crisis—Pan American Health Organization. Regional Office for the Americas of the World Health Organization, July 27. https://www.paho.org/en/news/27-7-2020-stronger-together-campaign-highlight-importance-psychosocial-support-during-times
7. Pelley L (2020) "We're all in this together": the phrase uniting Toronto in long, lonely battle against COVID-19. CBC News, March 29. https://www.cbc.ca/news/canada/toronto/we-re-all-in-this-together-the-phrase-uniting-toronto-in-long-lonely-battle-against-covid-19-1.5508850
8. Ramsden P (2020) How the pandemic changed social media and George Floyd's death created a collective conscience. The Conversation. http://theconversation.com/how-the-pandemic-changed-social-media-and-george-floyds-death-created-a-collective-conscience-140104
9. Pyszczynski T, Lockett M, Greenberg J, Solomon S (2022) Terror management theory and the COVID-19 pandemic. J Humanist Psychol 61(2):173–189
10. Oromoni A (2022) Tension between U of T law and disabled students raises a broader policy implication: lawyer. Law Times, June 8. https://www.lawtimesnews.com/resources/legal-education/tension-between-u-of-t-law-and-disabled-students-raises-a-broader-policy-implication-lawyer/367300
11. Walcott R (2019) The end of diversity. Public Cult 31(2):393–408. https://doi.org/10.1215/08992363-7286885
12. Xie Y, Bowe B, Al-Aly Z (2021) Burdens of post-acute sequelae of COVID-19 by severity of acute infection, demographics and health status. Nat Commun 12(1):6571
13. Azam TU, Colvin MM (2021) Representation of Black patients in heart failure clinical trials. Curr Opin Cardiol 36(3):329–334
14. Zuniga KB, Borno H, Chan JM et al (2020) The problem of underrepresentation: Black participants in lifestyle trials among patients with prostate cancer. J Racial Ethn Health Disparities 7(5):996–1002
15. Bergmans RS, Chambers-Peeple K, Aboul-Hassan D et al (2022) Opportunities to improve long COVID care: implications from semi-structured interviews with Black patients. Patient 15(6):715–728
16. Francis L (2020) Covid and the veil of ignorance. Philosophy Talk. Accessed 7 July 2022. https://www.philosophytalk.org/blog/covid-and-veil-ignorance
17. Rawls J (1971) A theory of justice, Original. Harvard University Press, Cambridge
18. Napier AD (2020) Rethinking vulnerability through Covid-19. Anthropol Today 36(3):1–2. https://doi.org/10.1111/1467-8322.12571
19. Gawthrop E (2022) Color of coronavirus: COVID-19 deaths analyzed by race and ethnicity. APM Research Lab. Accessed 15 Oct 2022. https://www.apmresearchlab.org/covid/deaths-by-race
20. Bell DA (1980) Brown v. Board of Education and the interest-convergence dilemma. Harvard Law Rev 93(3):518–533
21. U.S. Government Accountability Office (2022) K-12 education: student population has significantly diversified, but many schools remain divided along racial, ethnic, and economic lines. U.S. Government Accountability Office. Accessed 27 Nov 2022. https://www.gao.gov/products/gao-22-104737
22. Heckler N, Mackey J (2022) COVID's influence on Black Lives Matter: how interest convergence explains the 2020 call for equality and what that means for administrative racism. Public Integrity. 24(4–5):364–374
23. Goodman JD (2021) A year after 'Defund,' police departments get their money back. The New York Times, October 10. Accessed 8 Nov 2022. https://www.nytimes.com/2021/10/10/us/dallas-police-defund.html
24. Cascio J, Plant EA (2015) Prospective moral licensing: does anticipating doing good later allow you to be bad now? J Exp Soc Psychol 56:110–116
25. Ries NM, Thomson M (2020) Bioethics and universal vulnerability: exploring the ethics and practices of research participation. Med Law Rev 28(2):293–316

26. Mithani Z, Cooper J, Boyd JW (2021). Race, power, and COVID-19: a call for advocacy within bioethics. Am J Bioethics 21(2):11–18, citing Solórzano DG, Yosso TJ (2002) Critical race methodology: counter-storytelling as an analytical framework for education research. Qual Inquiry 8(1):23–44, p 25
27. ten Have H (2022) Bioethics after covid. In: ten Have H (ed) The covid-19 pandemic and global bioethics. Advancing global bioethics. Springer International Publishing, Heidelberg, pp 225–271
28. Ye Y, Zhang Q, Wei X, Cao Z, Yuan HY, Zeng DD (2022) Equitable access to COVID-19 vaccines makes a life-saving difference to all countries. Nat Hum Behav 6(2):207–216
29. Riaz MMA, Ahmad U, Mohan A et al (2021) Global impact of vaccine nationalism during COVID-19 pandemic. Trop Med Health 49(1):101
30. Emanuel EJ, Upshur REG, Smith MJ (2022) What covid has taught the world about ethics. N Engl J Med 387(17):1542–1545

Jane Cooper (she/her), BScH, MBE, JD, is a recent JD (Juris Doctor) graduate from the University of Toronto Faculty of Law. Prior to law school, she completed a Bachelor of Science Degree in Biology at Queen's University, and a Master of Bioethics at Harvard Medical School. She is passionate about environmental and health justice, equity, and the law. She is currently a judicial law clerk for the Superior Court of Ontario, and will clerk for the Federal Court of Canada for the 2024–25 term. This chapter was written in her personal capacity, and does not reflect the views of the Ontario Government.

Zamina Mithani (she/her), BSc, MBE, is a graduate of the Bioethics Masters program at Harvard Medical School and is currently completing her Doctor of Medicine (MD) degree at the University of British Columbia. She is passionate about the intersection of narrative, ethics, medical education, and clinical care. She has given workshops on issues on medical ethics across Canada and the US, and continues to advocate for justice in bioethics.

J. Wesley "Wes" Boyd (he/him), MD, Ph.D. is a professor of psychiatry and medical ethics at Baylor College of Medicine. He is also a Lecturer on Global Health and Social Medicine at Harvard Medical School. He obtained an MA in philosophy and a Ph.D. in religion and culture, along with his medical degree, at UNC Chapel Hill. He completed psychiatry residency at Cambridge Hospital and a fellowship in Medical Ethics at Harvard Medical School. He has taught extensively in the humanities, bioethics, human rights and psychiatry. His areas of interest include social justice, access to care, human rights, asylum and immigration, humanistic aspects of medicine, physician health and well-being, the pharmaceutical industry, mass incarceration, and substance use, and writes for both academic and lay audiences in these areas.

Index

S. Achella and C. Marazia (eds.), *Vulnerabilities*, Integrated Science 18,
https://doi.org/10.1007/978-3-031-39378-5

MIX
Papier aus verantwortungsvollen Quellen
Paper from responsible sources
FSC® C105338

If you have any concerns about our products,
you can contact us on
ProductSafety@springernature.com

In case Publisher is established outside the EU,
the EU authorized representative is:
Springer Nature Customer Service Center GmbH
Europaplatz 3, 69115 Heidelberg, Germany

Printed by Libri Plureos GmbH
in Hamburg, Germany